T0136840

Studies in Systems, Decision and Control

Volume 357

Series Editor

Janusz Kacprzyk, Systems Research Institute, Polish Academy of Sciences, Warsaw, Poland

The series "Studies in Systems, Decision and Control" (SSDC) covers both new developments and advances, as well as the state of the art, in the various areas of broadly perceived systems, decision making and control–quickly, up to date and with a high quality. The intent is to cover the theory, applications, and perspectives on the state of the art and future developments relevant to systems, decision making, control, complex processes and related areas, as embedded in the fields of engineering, computer science, physics, economics, social and life sciences, as well as the paradigms and methodologies behind them. The series contains monographs, textbooks, lecture notes and edited volumes in systems, decision making and control spanning the areas of Cyber-Physical Systems, Autonomous Systems, Sensor Networks, Control Systems, Energy Systems, Automotive Systems, Biological Systems, Vehicular Networking and Connected Vehicles, Aerospace Systems, Automation, Manufacturing, Smart Grids, Nonlinear Systems, Power Systems, Robotics, Social Systems, Economic Systems and other. Of particular value to both the contributors and the readership are the short publication timeframe and the world-wide distribution and exposure which enable both a wide and rapid dissemination of research output.

Indexed by SCOPUS, DBLP, WTI Frankfurt eG, zbMATH, SCImago.

All books published in the series are submitted for consideration in Web of Science.

More information about this series at http://www.springer.com/series/13304

Mojtaba Ahmadieh Khanesar · Okyay Kaynak ·
Erdal Kayacan

Sliding-Mode Fuzzy Controllers

 Springer

Mojtaba Ahmadieh Khanesar
Faculty of Engineering
University of Nottingham
Nottingham, UK

Okyay Kaynak
Department of Electrical
and Electronics Engineering
Bogazici University
Istanbul, Turkey

Erdal Kayacan
Department of Electrical and Computer
Engineering
Aarhus University
Aarhus, Denmark

ISSN 2198-4182   ISSN 2198-4190 (electronic)
Studies in Systems, Decision and Control
ISBN 978-3-030-69184-4   ISBN 978-3-030-69182-0 (eBook)
https://doi.org/10.1007/978-3-030-69182-0

This Springer imprint is published by the registered company Springer Nature Switzerland AG
The registered company address is: Gewerbestrasse 11, 6330 Cham, Switzerland

Preface

This book deals with motion control using sliding-mode fuzzy logic control. The sliding-mode fuzzy logic controller is a robust nonlinear control approach that has been successfully used in various applications. The main benefit of this controller is that it reduces tracking control to a bang–bang controller, directing states of the system to a sliding manifold, which defines the desired trajectory of the system and maintains them there. Such a controller is basically composed of a nominal control signal and a switching term to guarantee the robustness of the system against matched uncertainties. However, this controller suffers from chattering, high-frequency variations in the control signal, which may be caused by the switching term when there exist unmodeled dynamics and time delays in the control. Such a phenomenon excites the higher frequency dynamics of the system, gives rise to wears and tears, and results in the aging of the actuator, which, in turn, adds more uncertainties to the system. Sensitivity to noise is another issue that may be visited especially when states of the system are close to the sliding surface. Last but not the least, the nominal part of the sliding-mode control approach needs to be known *a priori*. Time variations in such a nominal part may disturb the controller. A fuzzy logic system is therefore designed to alleviate these issues by either using soft switching, rule-based control, or acting as an identifier to various parts of the nonlinear system. The diagram shown in Fig. 1 summarizes the existing sliding-mode fuzzy logic control approaches.

The main aim of this book is to provide enough material for those students with knowledge about ordinary differential equations and fuzzy logic controllers to master the design of sliding-mode fuzzy logic control methods. The chapters of this book are organized as follows:

Chapter 1 includes preliminary mathematics required to follow this book including Lie algebra, stability analysis, and Lyapunov theory. Some examples are presented to demonstrate how such mathematical tools are useful in the design of sliding-mode controllers.

Classical sliding-mode controllers are presented in Chap. 2. The main feature of the sliding-mode control approach is that it introduces the desired trajectory of the system in terms of a sliding mode whose order is less than the original system; further if the sliding mode is maintained, all of the system states converge to the desired point. The mathematical formulation of such a controller is presented and

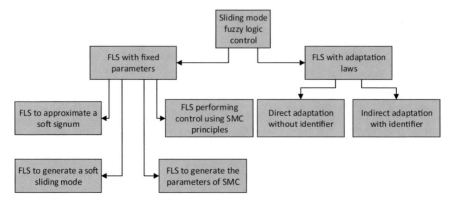

Fig. 1 Sliding-mode fuzzy logic control

its stability analysis is fully developed. Some examples are presented to demonstrate the applicability and robustness of the controller for systems modeled by ordinary differential equations.

Fuzzy logic as a method to represent human expert knowledge for its automatic usage in real-time systems is presented in Chap. 3. Interval type-2 fuzzy systems, which are a more promising method to deal with uncertainty and noise, are also presented in this section. The bottleneck of this system—its type reduction and defuzzifier—is presented and its existing solutions are demonstrated.

Rule-based sliding-mode fuzzy logic control is covered in Chap. 4. Fuzzy logic systems may be used to completely replace sliding-mode controllers, act instead of the sign function, shape the sliding mode in the system, or generate parameters of a sliding-mode controller.

Chapter 5 presents an adaptive sliding-mode fuzzy control using the gradient descent method and its second-order gradient alternative Levenberg–Marquardt. A cost function that includes a sliding mode is considered. The gradient descent method is then used to estimate the parameters of IT2FNN acting as a controller using such cost function.

In Chap. 6, a sliding-mode adaptive fuzzy logic controller with its parameter update rules being extracted from an appropriate Lyapunov function is covered. An interval type-2 fuzzy logic system as well as a type-1 fuzzy system are covered in this case, and comparisons are made in simulation to observe the superiority of the interval type-2 fuzzy systems over their type-1 counterparts.

Adaptive networked sliding-mode fuzzy logic control systems are covered in Chap. 7. Networked control systems suffer from time delays and packet losses, which may result in instability in the system. Padé approximation is used to deal with time delays in the system, which results in the design of an adaptive controller with closed-loop stability analysis. Parameter update rule modifications are added to compensate for the parameter instability in the system.

Teleoperation control is another application of sliding-mode fuzzy logic controllers; this controller may be used in hazardous and inaccessible areas to remotely control systems when it is difficult for a human expert to operate. This topic is covered in Chap. 8.

One of the main challenges facing the design of sliding-mode controllers is how to choose their parameters. Other than the energy of error, which is an important cost function to be considered, to dampen the chattering phenomenon, another cost function representing the chattering in a control signal is used. The resulting multi-objective optimization problem is solved using multi-objective approaches in Chap. 9.

We gratefully acknowledge Dr. Bibi Elham Fallah Tafti for her contributions in Chap. 9.

MATLAB files associated with this book can be downloaded from https://github.com/moji82/sliding_mode_fuzzy_control.

Nottingham, UK   Mojtaba Ahmadieh Khanesar
Istanbul, Turkey   Okyay Kaynak
Aarhus, Denmark   Erdal Kayacan

Contents

About the Authors

M. A. Khanesar is currently working as a postdoctoral research fellow within Advanced Manufacturing Technology Research Group at the University of Nottingham, UK. He has further held previous positions as a postdoctoral researcher at the Technical University of Denmark and an assistant professor at Semnan University, Iran. He is also a senior member of IEEE and serves as an academic editor for Complexity journal, a publication of John Wiley & Sons in collaboration with Hindawi Publishing Corporation. He received his B.Sc., M.S., and Ph.D. degrees in control engineering from K. N. Toosi University of Technology, Tehran, Iran, in 2005, 2007, and 2012, respectively. In 2010, he held a nine-month visiting student position at the Bogazici University, Istanbul, Turkey. His current research interests include manufacturing, robotics, machine learning, and control.

Okyay Kaynak received his B.Sc. (first-class honors) and Ph.D. degrees in electronic and electrical engineering from the University of Birmingham, U.K., in 1969 and 1972, respectively. From 1972 to 1979, he held various positions within the industry. In 1979, he joined the Department of Electrical and Electronics Engineering, Bogazici University, Istanbul, Turkey, where he is presently an Emeritus Professor. He also holds a "1000 Talents Program" Professor title at the University of Science and Technology, Beijing, China. He has held long-term (near to or more than a year) Visiting Professor/Scholar positions at various institutions in Japan, Germany, U.S., Singapore, and China. His current main research interest is in the broad field of intelligent control. He has authored three books and edited five and authored or coauthored more than 400 papers that have appeared in various journals, books, and conference proceedings. He has been in the editorial boards of several journals. Currently, he is the Editor-in-Chief of the Springer journal: Discover Artificial Intelligence. Additionally, the Springer volume "Recent Advances in Sliding Modes: From Control to Intelligent Mechatronics" is dedicated to him to commemorate his lifetime impactful research and scholarly achievements and outstanding services to the profession.

Dr. Kaynak is a fellow of IEEE. He has served on many committees of IEEE and was the president of IEEE Industrial Electronics Society during 2002–2003.

Erdal Kayacan received a Ph.D. degree in electrical and electronic engineering at Bogazici University, Istanbul, Turkey, in 2011. After finishing his postdoctoral research in University of Leuven (KU Leuven) in 2014, he worked in Nanyang Technological University (NTU), Singapore, at the School of Mechanical and Aerospace Engineering as an assistant professor for 4 years. Currently, he is pursuing his research at Aarhus University at the Department of Electrical and Computer Engineering as an associate professor, and he is the Director of Artificial Intelligence in Robotics Laboratory (AiRLab).

He has since published more than 140 peer-refereed book chapters, journals, and conference papers in model-based and model-free control, parameter and state estimation, computer vision, motion and path planning for robots. He has completed a number of research projects which have focused on the design and development of ground and aerial robotic systems, vision-based control techniques, and artificial intelligence. He is currently involved in a number EU projects, some of which are "Reliable AI for Marine Robotics" by Horizon 2020—H2020-MSCA-ITN-2020, European Union, and "Open Deep Learning toolkit for Robotics" by Robotics Core Technology ICT-10-2019-2020, European Union. Dr. Kayacan is co-writer of a course book "Fuzzy Neural Networks for Real Time Control Applications, 1st Edition Concepts, Modeling and Algorithms for Fast Learning". He is a Senior Member of Institute of Electrical and Electronics Engineers (IEEE) and members of Computational Intelligence Society and Robotics and Automation Society. Since January 1, 2017, he is an Associate Editor of IEEE Transactions on Fuzzy Systems (TFS) and IEEE/ASME Transactions on Mechatronics (TMECH). He is also the manager of Junior Reviewer Program in TMECH, which is intended to introduce young researchers in the mechatronics research community to the best practices in peer-reviewing of scientific publications under the guidance of Editorial Board members.

Abbreviations

BMM	Biglarbegian–Melek–Mendel
CAN	Central area network
COS TR	Center of set type reduction
DC	Direct current
DDC	Direct digital control
DOF	Degree of freedom
EAs	Evolutionary algorithms
EIASC	Enhanced iterative algorithm with a stop condition
EKM	Enhanced Karnik–Mendel
FEL	Feedback error learning
GA	Genetic algorithm
GD	Gradient descent
IASC	Iterative algorithm with a stop condition
IM	Induction motor
IT2FLS	Interval type-2 fuzzy logic system
IT2MFs	Interval type-2 fuzzy membership functions
KM	Karnik–Mendel
MOEA	Multi-objective evolutionary algorithm
MOPSO	Multi-Objective particle swarm optimization
NCS	Networked control system
NFE	Number of fitness evaluation
NNC	Neural network control
NSGA	Non-Dominated Sorting Genetic Algorithm
NT	Nie–Tan
PID	Proportional Integral Derivative
PSO	Particle swarm optimization
PWM	Pulse width modulation
SISO	Single-input–single-output
SMC	Sliding-mode control
SMFC	Sliding-mode fuzzy control
SPEA2	Strength Pareto Evolutionary Algorithm 2
T1FLS	Type-1 Fuzzy Logic System

T2MF	Type-2 MF
TCP	Transmission control protocol
TSFLC	Takagi–Sugeno fuzzy logic controller
TSFLS	Takagi–Sugeno fuzzy logic system
UDP	User datagram protocol

Chapter 1
Preliminaries

1.1 Introduction

It is often useful to replace a function (or a system) with its simpler version. Since linear system theory is very mature and has advanced tools to design a controller with respect to the design requirements, researchers prefer to deal with linearized versions of nonlinear systems. On the other hand, linearized models are acceptable when the system is operated around its linearization point. When systems are forced to operate far from their linearization points, where the nonlinear dynamics are excited, the performance of the system reduces drastically. For instance, where a linear controller for an unmanned aerial vehicle around its trim conditions works well, these controllers suffer when the aerial vehicle tries to perform aggressive maneuvers, i.e., large pitch and roll angles.

The feedback linearization method approaches the aforementioned problem differently, and is one of the main nonlinear control approaches in the literature. Instead of linearizing the system around a specific equilibrium point, a nonlinear feedback controller is designed, which eliminates the nonlinear terms. The result will be a simple linear system that does not have any nonlinear term in its model. Even though feedback linearization is a powerful technique for analysis and controller design for nonlinear systems, there are some challenges as well. First, the nonlinear terms in the model must be measured or estimated perfectly, and be fed into the controller. Any measurement error in this step will result in a mismatch between the real system and its corresponding linear model. Moreover, not all nonlinear systems are in a suitable form for feedback linearization. Hence, the following two fundamental questions must be answered: 1. What are the requirements for a system to be transformed to a form that can be linearized easily using a feedback controller? 2. What is the algebraic transform that can do this?

In this section, some mathematical tools, such as Lie algebra and the Lie bracket, are introduced and the necessary conditions to transform a system to the feedback linearization form are discussed.

© Springer Nature Switzerland AG 2021
M. Ahmadieh Khanesar et al., *Sliding-Mode Fuzzy Controllers*, Studies in Systems, Decision and Control 357, https://doi.org/10.1007/978-3-030-69182-0_1

1.2 Mathematical Tools

1.2.1 Lie Derivative

The Lie derivative be defined as follows:

$$L_f h(x) = \frac{\partial h(x)}{\partial x} f(x) \tag{1.1}$$

where $h(x) : R^n \to R$ and $f(x) : R^n \to R^n$ are two nonlinear functions of $x \in R^n$, and $\frac{\partial h(x)}{\partial x}$ represents the partial derivative of function $h(x)$ with respect to its input vector $x \in R^n$. The Lie derivative introduces the derivative of the scalar function $h(x)$ along the system $\dot{x} = f(x)$. Moreover, the second-order Lie derivatives are defined as follows:

$$L_f^2 h(x) = L_f L_f h(x) = \frac{\partial L_f h(x)}{\partial x} f(x) \tag{1.2}$$

Further, its higher-order Lie derivative is defined as follows:

$$L_f^n h(x) = L_f L_f^{n-1} h(x) = \frac{\partial L_f^{n-1} h(x)}{\partial x} f(x). \tag{1.3}$$

It is also possible to take the Lie derivative with respect to the two vector fields of $f(x) : R^n \to R^n$ and $g(x) : R^n \to R^n$, as follows:

$$L_g L_f h(x) = \frac{\partial L_f h(x)}{\partial x} g(x). \tag{1.4}$$

Example. The nonlinear vector functions $f(x)$ and $g(x)$ and the scalar function $h(x)$ are defined as follows:

$$h(x) = \frac{1}{2} x_1^2 + x_2 \tag{1.5}$$

$$f(x) = \begin{bmatrix} -x_1 \\ -2x_1 x_2 + 2(1 - x_1^2 - x_2^2) \end{bmatrix}, \quad g(x) = \begin{bmatrix} -x_1^2 \\ 1 - x_1^2 - x_2^2 \end{bmatrix}. \tag{1.6}$$

In this case, the following Lie derivatives are obtained:

$$L_f h(x) = \frac{\partial h}{\partial x} f(x)$$

$$= \begin{bmatrix} x_1 & 1 \end{bmatrix} \begin{bmatrix} -x_1 \\ -2x_1 x_2 + 2(1 - x_1^2 - x_2^2) \end{bmatrix}$$

$$= 2 - 2x_1 x_2 - 3x_1^2 - 2x_2^2 \tag{1.7}$$

and

$$L_g h(x) = \frac{\partial h}{\partial x} g(x)$$
$$= \begin{bmatrix} x_1 & 1 \end{bmatrix} \begin{bmatrix} -x_1^2 \\ 1 - x_1^2 - x_2^2 \end{bmatrix}$$
$$= -x_1^3 - x_1^2 - x_2^2 + 1 \qquad (1.8)$$

and

$$L_f^2 h(x) = \frac{\partial L_f h}{\partial x} f(x)$$
$$= \begin{bmatrix} -2x_2 - 6x_1 & -2x_1 - 4x_2 \end{bmatrix} \begin{bmatrix} -x_1 \\ -2x_1 x_2 + 2(1 - x_1^2 - x_2^2) \end{bmatrix}$$
$$= -4x_1^3 + 4x_1^2 x_2 + 6x_1^2 + 4x_1 x_2^2 + 2x_1 x_2 + 4x_1 + 8x_2^3 - 8x_2 \quad (1.9)$$

and

$$L_f L_g h(x) = \frac{\partial L_g h}{\partial x} f(x)$$
$$= \begin{bmatrix} -2x_1 - 3x_1^2 & -2x_2 \end{bmatrix} \begin{bmatrix} -x_1 \\ -2x_1 x_2 + 2(1 - x_1^2 - x_2^2) \end{bmatrix}$$
$$= 3x_1^3 + 4x_1^2 x_2 + 2x_1^2 + 4x_1 x_2^2 + 4x_2^3 - 4x_2. \qquad (1.10)$$

1.2.2 *Lie Bracket*

The Lie bracket is an operator which operates on two field vectors. This kind of derivative is frequently used in nonlinear controller design and is defined as follows:

$$[f, \ g](x) = \frac{\partial g}{\partial x} f(x) - \frac{\partial f}{\partial x} g(x). \qquad (1.11)$$

Example 1.1 Let the nonlinear vector functions $f(x)$ and $g(x)$ be as follows:

$$f(x) = \begin{bmatrix} -x_1 \\ -2x_1 + 2(1 - x_1 x_2 - x_2^2) \end{bmatrix}, \ g(x) = \begin{bmatrix} x_1^2 \\ 1 - x_2^2 \end{bmatrix}. \qquad (1.12)$$

The following Lie bracket is obtained:

Table 1.1 Lie algebra properties

$ad_f^0 g(x) = g(x)$	Zero-order Lie bracket
$ad_f^2 g(x) = [f, \ ad_f g] = [f, [f, g]]$	Second-order Lie bracket
$ad_f^i g(x) = [f, \ ad_f^{i-1} g]$	Higher-order Lie bracket
$[f, \ g] = -[g, \ f]$	Commutative property

$$[f, \ g](x) = \frac{\partial g}{\partial x} f(x) - \frac{\partial f}{\partial x} g(x)$$

$$= \begin{bmatrix} 2x_1 & 0 \\ 0 & -4x_2 \end{bmatrix} \begin{bmatrix} -x_1 \\ -2x_1 + 2(1 - x_1 x_2 - x_2^2) \end{bmatrix}$$

$$- \begin{bmatrix} -1 & 0 \\ -2 - 2x_2 & -2x_1 - 4x_2 \end{bmatrix} \begin{bmatrix} x_1^2 \\ 1 - x_2^2 \end{bmatrix}$$

$$= \begin{bmatrix} -x_1^2 \\ 2x_1^2 x_2 + 2x_1^2 + 6x_1 x_2^2 + 8x_1 x_2 + 2x_1 + 4x_2^3 - 4x_2 \end{bmatrix}. \quad (1.13)$$

Another notation that may be used for the Lie bracket is as follows:

$$[f, \ g](x) \equiv ad_f g(x). \quad (1.14)$$

Some basic properties of the Lie bracket are listed in Table 1.1.

1.2.3 Diffeomorphism

To transform a nonlinear dynamic system into a feedback linearization form, it is required to use a diffeomorphism transform to map the system states to a new space, in which the system can easily be linearized using a feedback linearization. A change in variable $z = T(x)$ is diffeomorphism in a neighborhood of origin if the inverse map $T^{-1}(.)$ exists and both $T(.)$ and $T^{-1}(.)$ are continuously differentiable.

Let the nonlinear system to be transformed be in the form of $\dot{x} = f(x)$. Consider a nonlinear diffeomorphism map as follows:

$$z = T(x). \quad (1.15)$$

The dynamics of the transformed system is obtained as follows:

$$\dot{z} = \frac{\partial T(x)}{\partial x} f(x) \bigg|_{x = T^{-1}(z)}. \quad (1.16)$$

1.2.4 *Change in Coordination*

If a system is not originally in a feedback linearization form, it is required to use change in coordinate to convert it to a feedback linearization form.

Consider the general nonlinear system as follows:

$$\dot{x} = f(x) + g(x)u \tag{1.17}$$

where $f(x) : R^n \rightarrow R^n$ and $g(x) : R^n \rightarrow R^n$ are nonlinear functions in the system. The change in coordination is chosen as $z = T(x)$, in which $T(x)$ is a diffeomorphism transformation. The dynamics of the system in the new coordinations must be determined. The time derivative of the parameter z is obtained as follows:

$$\dot{z} = \frac{\partial T}{\partial x}\left[f(x) + g(x)u\right] \tag{1.18}$$

which may further be written using Lie algebra in the following form.

$$\dot{z} = L_f T(x) + L_g T(x)u. \tag{1.19}$$

Since $T(x)$ is a diffeomorphism transformation, its inverse exists as follows:

$$x = T^{-1}(z). \tag{1.20}$$

Finally, the exchange of the states of the original system with the transformed states in the new coordinates is required.

In order to illustrate the change in the coordinate using an appropriate example, the following example is given.

Example 1.2 Consider a nonlinear dynamic system in the following form:

$$\dot{x} = \begin{bmatrix} 2 \\ -4x_1 \\ x_1^2 + x_3 \end{bmatrix} + \begin{bmatrix} 1 \\ -2x_1 \\ 8x_1x_2 + 2x_1 \end{bmatrix} u. \tag{1.21}$$

The diffeomorphism transform is considered to be as follows:

$$z = T(x) = \begin{bmatrix} 2x_1 \\ 4x_1^2 + 4x_2 \\ x_2^2 + x_3 \end{bmatrix} \tag{1.22}$$

whose inverse transform, which transforms the states from the secondary coordination to the original one, is as follows:

$$x = T^{-1}(z) = \begin{bmatrix} 0.5z_1 \\ z_2 - z_1^2 \\ z_3 - (z_2 - z_1^2)^2 \end{bmatrix}. \tag{1.23}$$

In this case, the time derivatives of the states in the new coordination are obtained as follows:

$$\dot{z} = \frac{\partial T(x)}{\partial x}[f(x) + g(x)u] \tag{1.24}$$

$$= \begin{bmatrix} 4 \\ 0 \\ x_1^2 - 8x_1x_2^2 + x_3 \end{bmatrix} + \begin{bmatrix} 2 \\ 0 \\ -4x_1x_2^2 + 8x_1x_2 + 2x_1 \end{bmatrix} u. \tag{1.25}$$

Considering (1.23), (1.24) can be rewritten as follows:

$$\dot{z} = \frac{\partial T(x)}{\partial x}[f(x) + g(x)u]$$

$$= \begin{bmatrix} 4 \\ 0 \\ z_3 - 4z_1(-z_1^2 + z_2)^2 - (-z_1^2 + z_2)^2 + z_1^2/4 \end{bmatrix}$$

$$+ \begin{bmatrix} 2 \\ 0 \\ z_1 - 2z_1(-z_1^2 + z_2)^2 + 4z_1(-z_1^2 + z_2) \end{bmatrix} u. \tag{1.26}$$

1.3 Input/State Linearization

It is highly desired to design a feedback control law with a coordination change that transforms a nonlinear dynamic system to a linear dynamic system. However, in order for a nonlinear system to be able to be converted to a linear one, some special conditions need to be satisfied.

Consider a nonlinear dynamic system as follows:

$$\dot{x} = Ax + B\phi(x)(u - v(x)) \tag{1.27}$$

where $x \in R^n$ is the system state vector, $A \in R^{n \times n}$, $B \in R^{n \times 1}$ are the matrices of the system, and $\phi(x) : R^n \to R$ and $v(x) : R^n \to R$ are two nonlinear functions of system states. It is possible to design a control signal $u = v(x) + \phi^{-1}(x)v$ to obtain a linear time-invariant system as follows:

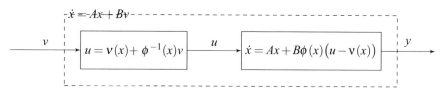

Fig. 1.1 Overall feedback linearized system

$$\dot{x} = Ax + Bv. \tag{1.28}$$

In this case, the control signal $u(t)$ is designed such that the system whose input is $v(t)$ is a linear system. The obtained linear time-invariant system can easily be controlled using a feedback control signal (see Fig. 1.1).

To make the system linear, it is highly desirable for the system to be in the form of (1.27). A more general class of nonlinear dynamic systems is considered to be as follows:

$$\dot{x} = f(x) + g(x)u \tag{1.29}$$

where $x \in R^n$, $f(x) : R^n \rightarrow R$, and $g(x) : R^n \rightarrow R$. In this case, it is possible that the original system is not in this form; however, there exists a diffeomorphism transformation $T(x) : R^n \rightarrow R^n$ that converts the original nonlinear dynamic system to the form of (1.27). The states of the transformed system are represented by z, which satisfy the following equation:

$$z = T(x). \tag{1.30}$$

Hence, the following equation is obtained:

$$\dot{z} = \frac{\partial T(x)}{\partial x}\dot{x}. \tag{1.31}$$

If (1.29) is replaced in (1.31), the following equation is obtained:

$$\dot{z} = \frac{\partial T(x)}{\partial x}\big(f(x) + g(x)u\big). \tag{1.32}$$

The following equation needs to be satisfied in order to make a system feedback linearizable:

$$\frac{\partial T(x)}{\partial x}\big(f(x) + g(x)u\big) = Az - B\phi(x)v(x) + B\phi(x)u. \tag{1.33}$$

It is possible to use a specific realization of the state-space form of the system. The state-space realizations of the system are not unique, and it is possible to use an invertible transformation to convert different realizations together. In this case, the controllable realization of the system is preferred whose matrices are of the following form:

$$
B = \begin{bmatrix} 0 \\ 0 \\ \vdots \\ 0 \\ 1 \end{bmatrix}, A = \begin{bmatrix} 0 & 1 & 0 & \cdots & 0 \\ 0 & 0 & 1 & \cdots & 0 \\ 0 & 0 & 0 & \ddots & 0 \\ 0 & 0 & 0 & \cdots & 1 \\ 0 & 0 & 0 & \cdots & 0 \end{bmatrix}.
\tag{1.34}
$$

The transformation $T(x)$ is considered to be as follows:

$$
T(x) = \begin{bmatrix} T_1(x) \\ T_2(x) \\ \vdots \\ T_n(x) \end{bmatrix}.
\tag{1.35}
$$

It follows from several equations that in order for a system to satisfy Eq. (1.33), it is required for the following equations to be satisfied:

$$
\frac{\partial T_1}{\partial x} f(x) = T_2(x)
$$

$$
\frac{\partial T_2}{\partial x} f(x) = T_3(x)
$$

$$
\vdots
$$

$$
\frac{\partial T_n}{\partial x} f(x) = -\phi(x)v(x)
\tag{1.36}
$$

and

$$
\frac{\partial T_1}{\partial x} g(x) = 0
$$

$$
\frac{\partial T_2}{\partial x} g(x) = 0
$$

$$
\vdots
$$

$$
\frac{\partial T_{n-1}}{\partial x} g(x) = 0
$$

$$
\frac{\partial T_n}{\partial x} g(x) = \phi(x).
\tag{1.37}
$$

Example 1.3 As an example to demonstrate how input/output feedback linearization is done, consider the magnetic levitation system. This system is composed of a magnetic system with a ball which must be suspended using the magnetic field of a

Table 1.2 Nomenclature corresponding to the magnetic levitation system

Symbol	Description	Numerical value
m	Mass of the ball	0.05 Kg
g	Gravitational acceleration	9.8 m/s^2
R	Resistance of the coil	1 Ω
L	Inductance of the coil	0.01 H

winding wire (see Fig. 1.2). It might also be considered as the simplified model of some magnetically levitated vehicles. The nonlinear dynamic model of this system is as follows:

$$\dot{x}_1 = x_2$$
$$\dot{x}_2 = g - \frac{Cx_3^2}{mx_1^2}$$
$$\dot{x}_3 = -\frac{R}{L}x_3 + \frac{2Cx_2x_3}{Lx_1^2} + \frac{1}{L}u \qquad (1.38)$$

with its parameters being defined as in Table 1.2.

According to the aforementioned conditions of input state linearization, it is highly desired that $T_1(x)$ is defined such that $\frac{\partial T_1}{\partial x}g(x) = 0$. In order to fulfill this requirement, the following condition must hold for $T_1(x)$:

$$\frac{\partial T_1}{\partial x}g(x) = \frac{1}{L}\frac{\partial T_1}{\partial x_3} = 0. \qquad (1.39)$$

This, in turn, requires that $T_1(x)$ be independent of the state of x_3. On the other hand, $T_2(x)$ is defined as follows:

$$T_2(x) = \frac{\partial T_1}{\partial x}f(x) = \frac{\partial T_1}{\partial x_1}x_2 + \frac{\partial T_1}{\partial x_2}\left(g - \frac{Cx_3^2}{mx_1^2}\right) + \frac{\partial T_1}{\partial x_2}\left(-\frac{R}{L}x_3 + \frac{2Cx_2x_3}{Lx_1^2}\right). \qquad (1.40)$$

In order to obtain a full feedback linearization form, it is also required that $\frac{\partial T_2}{\partial x}g(x) = 0$, which results in the following equation:

$$\frac{\partial T_2}{\partial x}g(x) = \frac{\partial T_2}{\partial x_3}g(x) - \frac{2Cx_3}{mx_1^2}\frac{\partial T_1}{\partial x_2} + \frac{\partial T_1}{\partial x_2}\left(-\frac{R}{L} + \frac{2Cx_2}{Lx_1^2}\right) = 0. \qquad (1.41)$$

In order to fulfill (1.41), it is possible to choose $T_1(x)$ such that it is independent of x_2, and T_2 such that it is independent of x_3. Considering these conditions, an appropriate selection for $T_1(x)$ may be $T_1(x) = x_1$, which results in the following selection for $T_2(x)$:

$$z_2 = T_2(x) = \frac{\partial T_1}{\partial x} f(x) = x_2. \tag{1.42}$$

Consequently, the third state of the system is obtained as follows:

$$z_3 = T_3(x) = \frac{\partial T_2}{\partial x} f(x) = g - \frac{C x_3^2}{m x_1^2}. \tag{1.43}$$

The dynamic model of the system in the new coordinate is obtained as follows:

$$\dot{z}_1 = z_2$$
$$\dot{z}_2 = z_3$$
$$\dot{z}_3 = \frac{\partial T_3}{\partial x}\big(f(x) + g(x)u\big). \tag{1.44}$$

By comparing the (1.44) with (1.33), $\phi(x)$ and $\nu(x)$ are obtained as follows:

$$-\phi(x)\nu(x) = \frac{\partial T_3}{\partial x} f(x) = \frac{2C}{m} - \frac{4C^2 x_3^2 x_2}{m L x_1^2} + \frac{2C R x_3^2}{m L x_1^2} \tag{1.45}$$

$$\phi(x)\frac{\partial T_3}{\partial x} g(x) = -\frac{2C x_3}{m L x_1^2}. \tag{1.46}$$

Hence, it is possible to choose the feedback linearizing control signal as being equal $u = \phi(x)^{-1}\nu + \nu(x)$ to make the original system of (1.38) completely linear (Fig. 1.1).

Example 1.4 Electro-hydraulic servo systems are applicable to a wide range of industrial applications where large inertia and torque loads require handling with high performance and speed [1, 2]. Typical applications of such systems are control of industrial robots, satellites, flight simulators, and many more.

The system states are x_1: hydro-motor angular velocity [rad/sec]; x_2: load pressure differential [Pa]; and x_3: valve displacement [m]. The dynamic equations of the system are as follows:

$$\dot{x}_1 = \frac{1}{J_t}\big(-B_m x_1 + q_m x_2 - q_m C_f P_s\big)$$

$$\dot{x}_2 = \frac{2\beta_e}{V_0}\left(-q_m x_1 - C_{im} x_2 + C_d W x_3 \sqrt{\frac{1}{\rho}(P_s - x_2)}\right)$$

$$\dot{x}_3 = \frac{1}{T_r}\left(-x_3 + \frac{K_r}{K_q} u\right)$$

$$y = x_1. \tag{1.47}$$

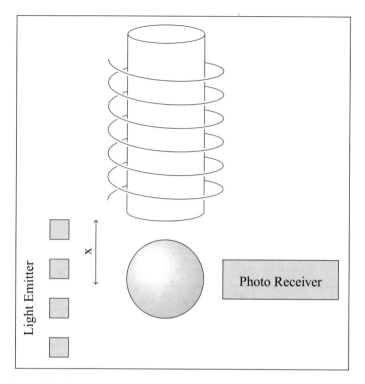

Fig. 1.2 Magnetic levitation system

The nomenclature of the symbols and the numerical values used in this study are given in Table 1.3.

In order to ease the notation of the nonlinear dynamic system of (1.47), its dynamic is written in the form of $\dot{x} = f(x) + g(x)u$ in which $f(x)$ and $g(x)$ are defined as follows:

$$f = \begin{bmatrix} 276.0 - 2.65 \times 10^{-4}x_2 - 0.0367x_1 \\ 3.55 \times 10^{11}x_3(1.18 \times 10^4 - 0.00118x_2)^{1/2} - 389.0x_2 - 1.85 \times 10^8 x_1 \\ -100.0x_3 \end{bmatrix} \quad (1.48)$$

$$g = \begin{bmatrix} 0 \\ 0 \\ 0.0084 \end{bmatrix}. \quad (1.49)$$

In order to show that the system is involutive, it is required that $[ad_f g, g] \in span\{ad_f g, g\}$. The vector field $ad_f g$ is obtained as follows:

Table 1.3 Nomenclature

Symbol	Description	Numerical value
J_t	Total inertia of the motor	$0.03\,\text{kgm}^2$
q_m	Volumetric displacement of the motor	$7.96 \times 10^{-7}\,\text{m}^3/\text{rad}$
B_m	Viscous damping coefficient	$1.1 \times 10^{-3}\,\text{Nms}$
C_f	Dimensionless internal friction coefficient	0.104
V_0	Average contained volume of each motor chamber	$1.2 \times 10^{-4}\,\text{m}^3$
β_e	Effective hulk modulus of the system	$1.391 \times 10^9\,\text{Pa}$
C_d	Discharge coefficient	0.61
C_{im}	Internal or cross-port leakage coefficient of the motor	$1.69 \times 10^{-11}\,\text{m}^3/\text{Pa.s}$
P_s	Supply pressure	$10^7\,\text{Pa}$
ρ	Oil density	$850\,\text{kg/m}^3$
T_r	Valve time constant	$0.01\,\text{s}$
K_r	Valve gain	$1.4 \times 10^{-4}\,\text{m}^3/\text{s.V}$
K_q	Valve flow gain	$1.66\,\text{m}^2/\text{s}$
W	Surface gradient	$8\pi \times 10^{-3}\,\text{m}$

$$ad_f g = -\frac{\partial f}{\partial x}g = \begin{bmatrix} 0 \\ -3.0 \times 10^9 (1.18 \times 10^4 - 0.00118x_2)^{1/2} \\ 0.843 \end{bmatrix}. \quad (1.50)$$

Furthermore, $[ad_f g, g]$ is given by the following equation:

$$[ad_f g, g] = -\frac{\partial ad_f g}{\partial x}g = \begin{bmatrix} 0 \\ 0 \\ 0 \end{bmatrix}. \quad (1.51)$$

As can be seen from (1.51), $[ad_f g, g] \in span\{ad_f g, g\}$, which means that $\Omega = [ad_f g, g]$ is involutive and the system is full state feedback linearizable. In order to find an appropriate transformation that transforms the system to the desired form of ((1.33)), the following requirement needs to be satisfied for $T_1(x)$:

$$\frac{\partial T_1(x)}{\partial x}g = \frac{\partial T_1(x)}{\partial x_3} = 0. \quad (1.52)$$

Hence, $T_1(x)$ must be independent of x_3. The function $T_2(x)$ is obtained as follows:

$$T_2(x) = L_f T_1(x) = \frac{\partial T_1(x)}{\partial x} f(x) = \frac{\partial T_1(x)}{\partial x_1} f_1(x) + \frac{\partial T_1(x)}{\partial x_2} f_2(x)$$

$$= (276.0 - 2.65 \times 10^{-4} x_2 - 0.0367 x_1) \frac{\partial T_1(x)}{\partial x_1}$$

$$+ (3.55 \times 10^{11} x_3 (1.18 \times 10^4 - 0.00118 x_2)^{1/2}$$

$$- 389.0 x_2 - 1.85 \times 10^8 x_1) \frac{\partial T_1(x)}{\partial x_2}. \tag{1.53}$$

It is required that $\frac{\partial T_2(x)}{\partial x} g = 0$, which results in the following equation:

$$\frac{\partial T_2(x)}{\partial x} g = \frac{\partial T_2(x)}{\partial x_3} = 0 \tag{1.54}$$

which further requires the following equation be satisfied:

$$\frac{\partial T_2(x)}{\partial x} g = \frac{\partial T_2(x)}{\partial x_3} = 3.55 \times 10^{11} (1.18 \times 10^4 - 0.00118 x_2)^{1/2} \frac{\partial T_1(x)}{\partial x_2} = 0. \tag{1.55}$$

Hence, it is required that

$$\frac{\partial T_1(x)}{\partial x_2} = 0. \tag{1.56}$$

In this case, in order to guarantee that (1.56) holds, T_1 must be selected to be independent of x_2 as well. Therefore, T_1 must be selected as a function of x_1. In this case, $T_1(x) = x_1$ is selected. This selection satisfies the following two conditions:

$$\frac{\partial T_1(x)}{\partial x} g = \frac{\partial T_2(x)}{\partial x} g = 0. \tag{1.57}$$

Hence, the first state in the new coordination is defined to be equal to $z_1 = T_1(x) = x_1$. Consequently, the second state in the new coordination is obtained as follows:

$$z_2 = T_2(x) = \frac{\partial T_1(x)}{\partial x} f(x) = 276.0 - 2.65 \times 10^{-4} x_2 - 0.0367 x_1. \tag{1.58}$$

It is required that the transformation to the new coordinate be invertible. Hence, the inverse transformation is obtained as follows:

$$x_2 = 1.0415 \times 10^6 - 138.4906 z_1 - 3.7736 \times 10^3 z_2. \tag{1.59}$$

We can proceed to obtain the third state of the system in the new coordinate, as follows:

$$z_3 = T_3(x) = \frac{\partial T_2(x)}{\partial x} f(x)$$

$$= [-0.0367 \quad -2.65 \times 10^{-4} \quad 0] \times$$

$$\begin{bmatrix} 276.0 - 2.65 \times 10^{-4}x_2 - 0.0367x_1 \\ 3.55 \times 10^{11}x_3(1.18 \times 10^4 - 0.00118x_2)^{1/2} - 389.0x_2 - 1.85 \times 10^8 x_1 \\ -100.0x_3 \end{bmatrix} \quad (1.60)$$

which results in the following equation between the system states in the newer coordination with respect to the original coordination of the system:

$$z_3 = -10.13 - 4.9555 \times 10^4 x_1 + 0.1031x_2 - 94075000x_3(1.18 \times 10^4 - 0.00118x_2)^{1/2}. \quad (1.61)$$

Finally, the nonlinear dynamics of the system in the newer coordination is obtained as follows:

$$\dot{z}_1 = z_2$$
$$\dot{z}_2 = z_3$$
$$\dot{z}_3 = \frac{\partial T_3(x)}{\partial x} f(x) + \frac{\partial T_2(x)}{\partial x} g(x)u. \quad (1.62)$$

On comparing (1.62) with (1.33), the following equations are obtained for $\phi(x)$ and $v(x)$:

$$-\phi(x)v(x) = \frac{\partial T_3(x)}{\partial x} f(x) = -4.9 \times 10^4 f_1 - 0.1f_2 - \frac{1.6 \times 10^8 x_3(0.01f_2 - 2.0x_2 + 2.0 \times 10^7)}{\sqrt{1.0 \times 10^7 - 1.0x_2}} \quad (1.63)$$

$$\phi(x) = \frac{\partial T_2(x)}{\partial x} g(x) = 2.7 \times 10^4 \sqrt{1.0 \times 10^7 - 1.0x_2}, \quad (1.64)$$

which causes the feedback linearizing control signal to be in the following form:

$$u = \frac{v}{\phi(x)} + v(x). \quad (1.65)$$

1.4   Input–Output Linearization

Let the single-input–single-output dynamic system be as follows:

$$\dot{x} = f(x) + g(x)u \quad (1.66)$$
$$y = h(x)$$

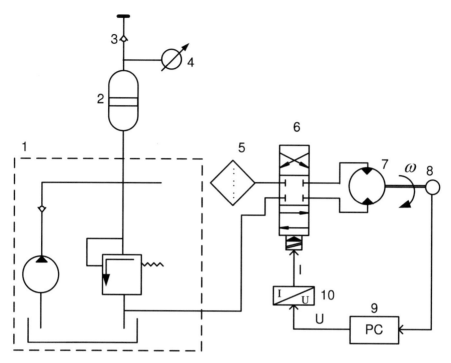

Fig. 1.3 Electro-Hydraulic Servo System. Reproduced from the copyright material source: M. Jovanovic, "Nonlinear control of an electrohydraulic velocity servosystem", *in Proceedings of the 2002 American Control Conference (IEEE Cat. No. CH37301)*, Vol. 1. IEEE, 2002, pp. 588–593 with the permission of the American Automatic Control Council (AACC)

where $f(x) : R^n \rightarrow R^n$, $g(x) : R^n \rightarrow R^n$, and $h(x) : R^n \rightarrow R^n$ are sufficiently smooth nonlinear functions in a domain $D \subset R$, with y being the system output.

$$\dot{y} = \frac{\partial y}{\partial x}[f(x) + g(x)u] = L_f h(x) + L_g h(x)u. \qquad (1.67)$$

Since it is required that the time derivative of the input signal u does not appear in the state-space dynamic of the system, if $L_g h(x) \neq 0$, no further time derivatives of the output are calculated. On the other hand, if $L_g h(x) = 0$, then the second derivative of the system output y with respect to time is calculated, which yields

$$\ddot{y} = \frac{\partial L_f h(x)}{\partial x}[f(x) + g(x)u] = L_f^2 h(x) + L_g L_f h(x)u. \qquad (1.68)$$

If $L_g L_f h(x) = 0$, the same process is repeated once more to obtain the following equation:

Fig. 1.4 The ball and beam
system

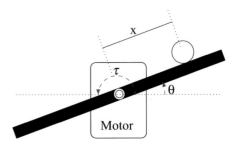

$$y^{(\rho)} = \frac{\partial L_f^{(\rho-1)} h(x)}{\partial x} \left[f(x) + g(x)u \right] = L_f^{(\rho)} h(x) + L_g L_f^{(\rho-1)} h(x)u \quad (1.69)$$

which requires the following condition to be satisfied:

$$L_g L_f^{i-1} h(x) = 0, \quad i = 1, 2, \ldots, \rho - 1; \ L_g L_f^{\rho-1} h(x) \neq 0. \quad (1.70)$$

Hence, this system is input–output feedback linearizable using the following control
signal:

$$u = \frac{1}{L_g L_f^{\rho-1} h(x)} \left[-L_f^{\rho} h(x) + v \right] \quad (1.71)$$

which reduces the system to an integral system of order ρ as follows:

$$y^{\rho} = v. \quad (1.72)$$

Moreover, if ρ is equal to n, then the system is full-state linearizable. If $\rho < n$, the
system has zero dynamics, which is more difficult to control (Fig. 1.4).

Example 1.5 The input–output feedback linearization is implemented on a ball and
beam system. This system is a highly used benchmark example in control engineering.

The nonlinear dynamic equation of the system is as follows:

$$\begin{bmatrix} \dot{x}_1 \\ \dot{x}_2 \\ \dot{x}_3 \\ \dot{x}_4 \end{bmatrix} = \begin{bmatrix} x_2 \\ B\left(x_1 x_4^2 - g sin(x_3)\right) \\ x_4 \\ 0 \end{bmatrix} + \begin{bmatrix} 0 \\ 0 \\ 0 \\ 1 \end{bmatrix} = f(x) + g(x)u$$

$$y = x_1 = h(x). \quad (1.73)$$

Hence, the time derivative of the system output is obtained as follows:

$$\dot{y} = L_f h(x) \tag{1.74}$$

$$= [1, \ 0, \ 0, \ 0] \begin{bmatrix} x_2 \\ B(x_1 x_4^2 - g\sin(x_3)) \\ x_4 \\ 0 \end{bmatrix} = x_2. \tag{1.75}$$

Since $L_g h(x) = 0$, it is possible to continue to take the time derivative from the system output. The second-order derivative of the system output is obtained as follows:

$$\ddot{y} = L_f^2 h(x) =$$

$$= [0, \ 1, \ 0, \ 0] \begin{bmatrix} x_2 \\ B(x_1 x_4^2 - g\sin(x_3)) \\ x_4 \\ 0 \end{bmatrix} = B(x_1 x_4^2 - g\sin(x_3)). \tag{1.76}$$

In this case, since $L_g L_f h(x) = 0$, it is possible to continue to take the third-order time derivative from the system output as follows:

$$y^{(3)} = L_f^3 h(x) + L_g L_f^2(x)u$$

$$= [Bx_4^2, \ 0, \ -Bg\cos(x_3), \ 2x_1 x_4] \begin{bmatrix} x_2 \\ B(x_1 x_4^2 - g\sin(x_3)) \\ x_4 \\ 0 \end{bmatrix}$$

$$+ [Bx_4^2, \ 0, \ -Bg\cos(x_3), \ 2x_1 x_4] \begin{bmatrix} 0 \\ 0 \\ 0 \\ 1 \end{bmatrix}$$

$$= B(x_2 x_4^2 - g x_4 \cos(x_3)) + 2Bx_1 x_4 u$$

$$\tag{1.77}$$

Hence, the overall nonlinear dynamic of the system can be transformed to the input–output linearization form. The control signal that makes the overall system linear is as follows:

$$u = \frac{x_2 x_4^2 - g x_4 \cos(x_3) + v}{x_1 x_4} \tag{1.78}$$

which transforms the system to the following linear system:

$$\dot{z}_1 = z_2$$
$$\dot{z}_2 = z_3$$
$$\dot{z}_3 = z_4$$
$$\dot{z}_4 = v \tag{1.79}$$

whose poles are all on the origin.

1.5 Definition of Stability

Let a nonlinear autonomous system be in the following form:

$$\dot{x} = f(x) \tag{1.80}$$

where $f(x) : R^n \rightarrow R^n$. The equilibrium points of the system are obtained by solving the following nonlinear equation:

$$f(x) = 0. \tag{1.81}$$

Consider the equilibrium point of the system to be located at $x = 0$. The equilibrium point of the system (1.80) is stable, if for each $\varepsilon > 0$, there exists $\delta(\varepsilon) > 0$ such that

$$\|x(0)\| < \delta \Rightarrow \|x(t)\| < \epsilon, \quad \forall t \geq 0. \tag{1.82}$$

Hence, stability means that it is possible to choose the initial conditions of the system such that the system states remain bounded in a neighborhood near the equilibrium point near *zero* whose radius can be chosen as desired. The same system is unstable if it is not stable.

However, a more tighter definition of stability belongs to the asymptotic stability of the system, in which it is required that the system states converge to *zero*. A system is called asymptotically stable if it is possible to choose $\delta > 0$ such that [3]

$$\|x(0)\| < \delta \Rightarrow \lim_{t \to \infty} x(t) = 0. \tag{1.83}$$

The concept of stability is visually presented in Fig. 1.5. In this figure, the regions S_δ and S_ϵ are defined as follows:

$$S_\delta = \{x \mid \|x\| < \delta\}$$
$$S_\epsilon = \{x \mid \|x\| < \epsilon\}. \tag{1.84}$$

Furthermore, $x(0)$ represents the initial condition of the states of the system. While Curve 1 represents a possible asymptotically stable trajectory for the system, curve 2

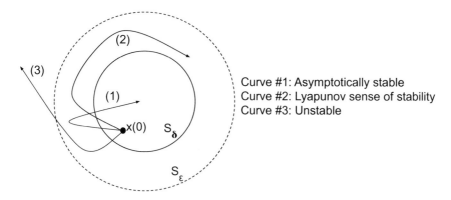

Curve #1: Asymptotically stable
Curve #2: Lyapunov sense of stability
Curve #3: Unstable

Fig. 1.5 Stability notion

demonstrates another stable trajectory that is not asymptotic. This is mainly because to fulfill asymptotic stability requirements, it is not enough for the system states to stay in ϵ-neighborhood, S_ϵ of the origin, but rather, it is required that the states of the system to converge to the origin. Trajectory 3 illustrates an unstable trajectory as the system states opt out of ϵ-neighborhood, S_ϵ as time goes. Hence, a system with a trajectory similar to Trajectory 3 is said to be unstable.

1.6 Stability Analysis

In order to prove the stability of a nonlinear ordinary differential equation, the Lyapunov stability theory is the most powerful and easily applicable method [3]. In order to investigate the stability of a system using this method, a positive function of the states on the trajectory of the differential equation is taken into account which is called the Lyapunov function. A Lyapunov function decreases along every point on the trajectory of the ordinary differential equation.

Let the ordinary differential equation of the system be of the following form:

$$\dot{x} = F(x) \tag{1.85}$$

where $F(x) : \mathbb{R}^n \to \mathbb{R}^n$ and $x \in \mathbb{R}^n$ is the state vector of the system. In this case, the following theorem holds.

Theorem 1.1 (The Stability of Continuous Time Systems [4]) *Let $x = 0$ be an equilibrium point and $D \in \mathbb{R}^n$ be a domain containing $x = 0$. Let $V : D \to \mathbb{R}$ be a continually differentiable function such that*

$$V(0) = 0, \quad and \quad V(x) > 0 \ in \ D - \{0\} \tag{1.86}$$

and

$$\dot{V} \le 0 \ in \ D, \tag{1.87}$$

then $x = 0$ *is stable. Moreover, if*

$$\dot{V} < 0 \ in \ D - \{0\}, \tag{1.88}$$

then $x = 0$ *is asymptotically stable.*

It is also possible to investigate the stability of a discrete-time difference equation using the Lyapunov theory.

Example 1.6 Consider a nonlinear dynamic system described by ordinary differential equations as follows:

$$\dot{x}_1(t) = x_2(t) - 0.1x_1(t)$$
$$\dot{x}_2(t) = -x1(t) - x_2(t) - x2(t)|x1(t)|. \tag{1.89}$$

The origin is globally asymptotically stable. In order to prove this claim, the following Lyapunov function is considered:

$$V = \frac{1}{2}x_1^2(t) + \frac{1}{2}x_2^2(t). \tag{1.90}$$

The time derivative of this Lyapunov function is obtained as follows:

$$\dot{V} = x_1(t)\dot{x}_1(t) + x_2(t)\dot{x}_2(t). \tag{1.91}$$

Substituting the dynamics of the system (1.89) in the time derivative of the Lyapunov function (1.91), we have the following equation:

$$\dot{V} = -0.1x_1^2(t) - x_2^2(t) - x_2^2(t)|x_1(t)| \tag{1.92}$$

which is negative-definite within the real number domain R^2. The trajectories of this system are shown in Fig. 1.6.

Example 1.7 Consider a nonlinear dynamic system described by ordinary differential equations as follows:

$$\dot{x}_1(t) = x_2(t)$$
$$\dot{x}_2(t) = -x_1(t) + x_2(t)\,|x_2(t)|\,. \tag{1.93}$$

The trajectories of this system are shown in Fig. 1.7, which clearly shows that this system is nonlinear.

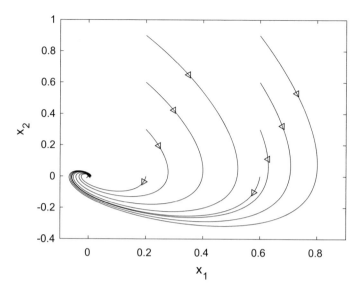

Fig. 1.6   Trajectories of the system in Example 1.6 from various initial conditions

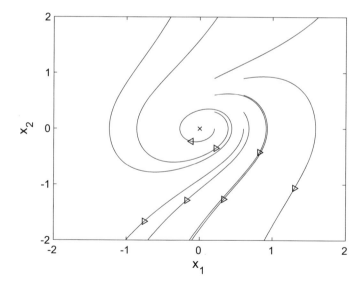

Fig. 1.7   Trajectories of the system in Example 1.7 from various initial conditions

Acknowledgements The authors gratefully acknowledge the permission granted by Prof. B. Wayne Bequette on behalf of the American Control Conference to reproduce the copyrighted material in this book: Fig. 3 from [5]. The authors gratefully acknowledge Prof. Mihailo Jovanovic, the author of the paper [5], for his correspondence for this figure.

References

1. Merritt, H.E.: Hydraulic Control Systems. Wiley, Hoboken (1967)
2. Watton, J.: Fluid Power Systems. Prentice Hall, Upper Saddle River (1989)
3. Slotine, J.-J.E., Li, W. et al.: Applied Nonlinear Control, vol. 199, no. 1. Prentice Hall, Englewood Cliffs (1991)
4. Khalil, H.: Nonlinear Control, Global Edition. Pearson Education Limited, London (2015)
5. Jovanovic, M.: Nonlinear control of an electrohydraulic velocity servosystem. In: Proceedings of the 2002 American Control Conference (IEEE Cat. No. CH37301), vol. 1, pp. 588–593. IEEE (2002)

Chapter 2
Classical Sliding-Mode Controllers

2.1 Introduction

Lack of an imprecise nonlinear model of real-time systems is inevitable due to the several simplifications that have been made, the neglected frictions, the dead-zones, and saturation. Moreover, in the case of having a perfect system model, uncertainty and variations in parameters due to aging, variations in environmental parameters such as temperature and humidity, and similar varying parameters do naturally exist.

One of the most well-known nonlinear control design tools to deal with uncertainty is the use of sliding-mode controllers, in which, the desired behavior of the system is defined in terms of a sliding manifold. The sliding manifold is stable, and the controller is designed to push system states to this manifold and maintain them on it. The convergence of the sliding manifold to zero is usually designed to happen in finite time, which necessitates the use of a switching function, the use of which improves the robustness of the system. However, a switching function may cause high-frequency oscillations in the control signal-chattering-and makes the system sensitive to noise [9].

The chattering in the control signal is due to the existence of uncertainty in the nonlinear model and the high-gain control action designed to guarantee the robustness in the presence of such uncertainties. The non-ideal and delayed switching is another cause of chattering. Chattering phenomenon makes the practical implementation of a sliding-mode controller very difficult—sometimes impossible. Moreover, it may lessen the durability of actuators used considerably. The excitation of high-frequency unmodeled dynamics of the system may cause instability.

Another disadvantage is sensitivity to noise. When the value of the sliding manifold is so close to zero, a small value of noise with a different sign than that of the sliding manifold may change the sign of the high-gain term of the sliding-mode controller.

© Springer Nature Switzerland AG 2021
M. Ahmadieh Khanesar et al., *Sliding-Mode Fuzzy Controllers*, Studies in Systems, Decision and Control 357, https://doi.org/10.1007/978-3-030-69182-0_2

The system controlled by a sliding-mode controller may exhibit different modes. Initially, system states may be outside of the sliding manifold. In this mode, the responsibility of a sliding-mode controller is to push system states toward the sliding manifold. This mode of operation is called the *reaching mode*. On the other hand, if the system states converge to the sliding manifold, they must be maintained on this surface. The latter is the so-called *sliding-mode*. Since the dynamic of the sliding manifold is a stable differential equation of the error, if the system states are kept on the sliding manifold, the error converges to zero.

In the sliding-mode controller design, the dynamic behavior of the system is defined by the sliding manifold. Hence, the design of a sliding mode is very important, requires trial and error, and may be done using intelligent optimization techniques. An integral type of sliding manifold exists in literature, which may result in a better tracking response with zero steady-state error. Time-varying and adaptive sliding surfaces also exist in literature, which may improve performance and lessen oscillations in the control signals.

Various modifications are made to the sliding-mode controllers to lessen the chattering phenomenon. An example of such an approach is the use of an approximation of the signum function, $\frac{s}{|s|+\delta}$. It is also possible to use the saturation function or the power law method to reduce chattering. Although these approaches lessen chattering considerably, they may disturb the robustness of the system. For instance, if the saturation function is used, the sliding mode converges to a small neighborhood of *zero* instead of an exact *zero* value.

Different adaptive sliding-mode approaches are also available in the literature. It may be possible to use an adaptive gain for the switching function, which avoids control gains that are too conservative. In other words, using the adaptive gain for the switching function makes it possible to use a smaller gain for the switching function, which may grow larger during control and become sufficiently high to maintain the robustness of the system. The unknown parameters of the system may also be determined during the control using an appropriate adaptation law.

In the design of terminal sliding-mode controllers, the sliding surface is designed such that its finite-time convergence to zero guarantees the finite-time convergence of the tracking error to zero as well. While in normal SMC, system states converge to zero with a dynamic defined by the sliding surface exponentially in infinite time, and the finite-time convergence of the states of the system that occurs in the terminal sliding mode is more desired. It is further possible to find the upper bound of the time that it takes for system states to converge to zero, which can be manipulated by appropriate selection of controller parameters.

It is also possible to use backstepping to design sliding-mode controllers for a more general class of nonlinear systems. While in the case of normal SMC, matched uncertainties can be compensated using the high-gain control signals, backstepping control method can effectively deal with some sort of mismatched uncertain nonlinearities.

2.2 SMC of Second-Order Nonlinear Systems

2.2.1 Constant Control Signal Coefficient Case

Let the second-order nonlinear dynamic system be as follows:

$$\begin{aligned} \dot{x}_1 &= x_2 \\ \dot{x}_2 &= f(x_1, x_2) + u, \end{aligned} \tag{2.1}$$

where $f(x_1, x_2)$ is a nonlinear scalar function of system states. It is assumed that $f(x_1, x_2)$ is partially known and has the following form:

$$f(x_1, x_2) = f_n(x_1, x_2) + \Delta f(x_1, x_2), \tag{2.2}$$

where $f_n(x_1, x_2)$ is the known part of $f(x)$ and $\Delta f(x)$ is the uncertain part of the function f, which satisfies $|\Delta F(x)| < F$. It is expected from the sliding-mode controller to drive system states to the sliding manifold and maintain sliding motion. The sliding manifold considered for the second-order differential equations for the system in (2.1) is as follows:

$$s = \lambda x_1 + x_2, \quad \lambda > 0, \tag{2.3}$$

where s is called the sliding manifold. Suppose that full sliding behavior is achieved and $s(t) = 0, \ \forall t > 0$. In this case, the solution to the differential equation in (2.3) is obtained as follows:

$$x_1(t) = x_1(0)e^{-\lambda t}, \tag{2.4}$$

where $x_1(0)$ represents the initial condition of system state. This equation shows that $x_1(t)$ tends to go to zero exponentially in infinite time. The parameter λ is a design parameter, which can be used to define the desired behavior of the system. This parameter plays an important role in the convergence speed of system states toward *zero*. Moreover, it can be seen from (2.4) that although the system is a second-order one, the desired motion behavior of the system is of first order. It is to be noted that the slope of this sliding surface is adjusted by λ, which determines how fast system states can converge to *zero*.

The time derivative of the sliding manifold is obtained as follows:

$$\dot{s} = \lambda \dot{x}_1 + \dot{x}_2 = \lambda x_2 + f(x) + u. \tag{2.5}$$

In order to prove the stability of the system, the following Lyapunov function is considered:

$$V = \frac{1}{2}s^2. \tag{2.6}$$

The control signal is designed such that the time derivative of the candidate Lyapunov function satisfies the following inequality:

$$s\dot{s} \le -\eta|s|. \tag{2.7}$$

The solution to the differential equation of $s\dot{s} = -\eta|s|$ is as follows:

$$s(t) = \begin{cases} s(0) - \eta.t.sign(s(0)), & t < \frac{s(0)}{\eta} \\ 0, & otherwise, \end{cases} \tag{2.8}$$

which guarantees that the sliding surface converges to zero in finite time, the value of which is equal to $\frac{s(0)}{\eta}$. In order to obtain the control signal, such that (2.7) is guaranteed, the following stages must be followed.

In the initial stage, the time derivative of the Lyapunov function is obtained as follows:

$$\dot{V} = s\dot{s}. \tag{2.9}$$

By replacing $\dot{s}$ as in (2.5) in (2.9), one obtains the following:

$$\dot{V} = s(\lambda x_2 + f(x_1, x_2) + u), \tag{2.10}$$

Further by using the fact that $f(x_1, x_2)$ has a nominal part and an uncertain part as in (2.2), we have the following equation for the time derivative of the Lyapunov function:

$$\dot{V} = s(\lambda x_2 + f_n(x_1, x_2) + \Delta f(x) + u). \tag{2.11}$$

The control signal u is assumed to have two parts as follows:

$$u = u_n + u_r, \tag{2.12}$$

where u_n represents a part of the control signal, which cancels the known part of the nonlinear function $f(x_1, x_2)$. This part of the control signal is called the equivalent control signal. The control signal also has another part, u_r, which is a term responsible for providing the robustness of system in the presence of $\Delta f(x)$. The signal, u_n, is taken as follows:

$$u_n = -\lambda x_2 - f_n(x_1, x_2). \tag{2.13}$$

The usage of the control signal (2.13), in the time derivative of the Lyapunov function in (2.11), results in the following form for the time derivative of the Lyapunov function:

$$\dot{V} = s(\Delta f(x) + u_r). \tag{2.14}$$

In order to have a stable Lyapunov function with finite time convergence to zero, the following equation must be satisfied:

$$\dot{V} \leq -\eta|s|, \tag{2.15}$$

and hence,

$$s(\Delta f(x) + u_r) < -\eta|s|, \tag{2.16}$$

where u_r is taken as follows:

$$u_r = -K sgn(s), \tag{2.17}$$

where $sgn(s)$ is the *Signum* function defined as follows:

$$sgn(s) = \begin{cases} 1, & s > 0 \\ -1 & s < 0 \\ 0 & s = 0. \end{cases} \tag{2.18}$$

By applying the control signal u_r as in (2.17)–(2.16), the following equation is obtained:

$$|s||\Delta f(x)| - K|s| < -\eta|s|, \tag{2.19}$$

which can be further manipulated using the fact that $|\Delta F(x)| < F$ as follows:

$$- K|s| < -\eta|s| - F|s|, \tag{2.20}$$

or equivalently,

$$\eta + F < K. \tag{2.21}$$

Hence, SMC signal, which guarantees the stability of system, is concluded as follows:

$$u = -\lambda x_2 - f_n(x) - K sgn(s) \tag{2.22}$$

A typical phase portrait of a second-order nonlinear dynamic system and its sliding manifold is depicted in Fig. 2.1. As can be seen from this figure, the initial states of the system are not necessarily on the sliding surface and it takes some time for them to hit the sliding surface. This mode is called reaching mode. However, when

Fig. 2.1 A typical phase
portrait under SMC

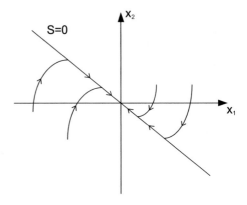

system states hit the sliding surface, the switching signal forces them on the sliding
manifold. This mode is called the sliding mode.

2.2.2 Nonlinear System with a Function as the Gain of Control Signal

Let us consider a more general class, in which,

$$\ddot{x}_1 = f(x) + g(x)u, \tag{2.23}$$

where $f(x) : R^2 \to R$ and $g(x) : R^2 \to R$ are nonlinear functions of system states
that are partially known as follows:

$$f(x) = f_n(x) + \Delta f(x), \;\; g(x) = g_n(x) + \Delta g(x), \tag{2.24}$$

where $f_n(x)$ is the known part of $f(x)$ and $\Delta f(x)$ is the uncertain part of the function
f that satisfies $|\Delta f(x)| < F$. Moreover, $g_n(x)$ is the known part of $g(x)$ with its
uncertain part being $\Delta g(x)$. Furthermore, it is assumed that $0 < g_{min} < g(x)$. The
state-space model of this system is as follows:

$$\begin{aligned} \dot{x}_1 &= x_2 \\ \dot{x}_2 &= f(x) + g(x)u. \end{aligned} \tag{2.25}$$

The sliding line considered for the system is as follows:

$$s = \dot{e} + \lambda e, \;\; \lambda > 0, \tag{2.26}$$

where $e = r - x$. Similar to the previous case, if the full sliding mode is obtained
and $s = 0$, then the dynamic behavior of the system becomes $e(t) = e(0)e^{-\lambda t}$, which

guarantees the convergence of $e(t)$ to zero. Hence, λ defines the desired behavior of the system. It is to be noted that although the system is a second-order nonlinear system, its tracking problem is reduced to force the sliding line to zero. The time derivative of the sliding line is obtained as follows:

$$\dot{s} = \ddot{r} - \ddot{x} + \lambda(\dot{r} - \dot{x}). \tag{2.27}$$

Consider the control signal u with two parts u_n and u_r, where u_n is the equivalent control signal that compensates the known parts of the dynamic model of the system, and u_r is the term that guarantees the finite-time convergence of the sliding line to zero,

$$u = u_n + u_r, \tag{2.28}$$

where the equivalent control signal is defined as follows:

$$u_n = \frac{1}{g_n}\left(-f_n + \ddot{r} + \lambda\dot{e}\right). \tag{2.29}$$

The control signal u_n satisfies the following equation:

$$g_n u_n + f_n - \ddot{r} - \lambda\dot{e} = 0. \tag{2.30}$$

If (2.30) is combined with (2.27), the following equation is obtained:

$$\begin{aligned}\dot{s} &= f_n - f + g_n u_n - g_n u_n - g u_r \\ &= \Delta f + \Delta g u_n - g u_r < -\eta|s|. \end{aligned} \tag{2.31}$$

In order to fulfill the inequality of (2.31), u_r is selected as follows:

$$u_r = \frac{K}{g_{min}}sgn(s). \tag{2.32}$$

The Lyapunov function considered for the system is selected as $V = \frac{1}{2}s^2$. Hence, its time derivative is obtained as $\dot{V} = s\dot{s}$. By applying (2.32)–(2.31) and considering the fact that $\dot{V} = s\dot{s}$, the following equation is obtained:

$$\dot{V} = s\dot{s} \leq F|s| + Gu_n|s| - K|s| \leq -\eta|s|. \tag{2.33}$$

In order to satisfy this inequality, K can be selected as $F + G|u_n| + \eta \leq K$.

Example 2.1 The tracking and balancing problem of an inverted pendulum on a cart (see Fig. 2.2) is considered in this example. The dynamic equations of motion governing the system are as follows [2]:

Fig. 2.2 The inverted
pendulum system

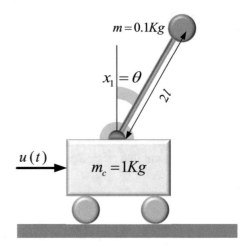

$$\dot{x}_1(t) = x_2(t)$$

$$\dot{x}_2(t) = \frac{gsin(x_1(t)) - mlax_2^2(t)cos(x_1(t))sin(x_1(t))}{l(\frac{4}{3} - macos^2(x_1(t)))}$$

$$+ \frac{acos(x_1(t))}{l(\frac{4}{3} - macos^2x_1(t))}u(t) + d(t), \tag{2.34}$$

where x_1 represents the angle of the pendulum, zero being its value when the pendulum is straight up. The signal $d(t)$ is the bounded external disturbance that is
considered to satisfy $|d(t)| < 0.2$. The angular velocity is denoted by x_2, g is the
gravity constant, $m = 0.1$ kg and $m_c = 1$ kg are the masses of the pendulum and the
cart, respectively, $a = 1/(m + m_c)$, and u is the force applied to the cart. g, the gravity constant, and l are considered as the partially known parameters. The nominal
values of g and l are considered as to be equal to 9.8 m/s^2 and 1 m, respectively.

The error signal for the system is defined as the difference between the reference
signal $r(t)$ and the system state $x_1(t)$ as $e(t) = x_1(t) - r(t)$. The sliding surface that
is considered to control this system is taken as $s = \dot{e} + e$. The equivalent control
signal u_n is obtained as follows:

$$u_n = \frac{1}{acos(x_1(t))}\left(-gsin(x_1(t)) + mlax_2^2(t)cos(x_1(t))sin(x_1(t))\right). \tag{2.35}$$

The term to guarantee the robustness of the system in the presence of uncertain
parameters and disturbances is considered to be as follows:

$$u_r = -Ksgn(s), \tag{2.36}$$

where K is selected as to be equal to 3.

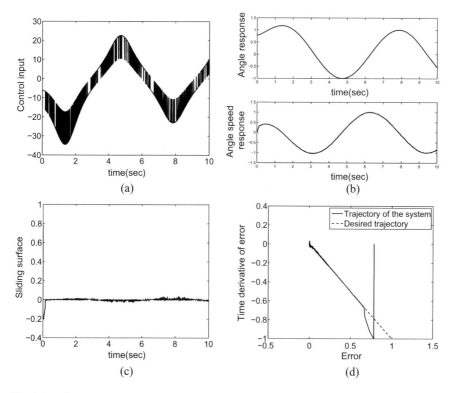

Fig. 2.3 a Control signal **b** Tracking response of the system **c** Sliding surface. **d** Trajectory of system

It is possible to use the *saturation* function instead of the *sign* function. This replacement may lessen chattering in the control signal. The *saturation* function is defined as follows:

$$sat(s, \phi) = \begin{cases} \frac{s}{\phi} & , |s| < \phi \\ sgn(s) & , \phi \leq |s|, \end{cases} \tag{2.37}$$

where ϕ is a design parameter that defines the width of the linearly behaved region of the *saturation* function. The results obtained when ϕ is selected as being equal to *zero* are illustrated in Fig. 2.3. As can be seen from this figure, the control signal is highly oscillating. In order to reduce oscillations in the next experiment, ϕ is selected as being equal to 0.1. The results are depicted in Fig. 2.4. As can be seen from this figure, the chattering in the control signal is highly reduced while maintaining the performance of system. However, it should be noted that when a saturation function is used instead of the signum function, the same Lyapunov function can only guarantee that the sliding manifold converges to a small neighborhood of zero and its convergence to zero cannot be guaranteed.

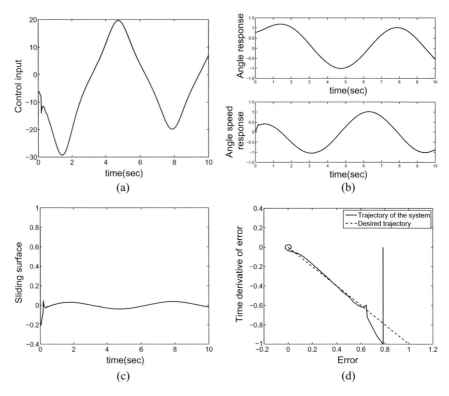

Fig. 2.4 a The control signal **b** The tracking response of the system **c** The sliding surface. **d** The trajectory of the system

2.3 Integral Sliding Surface

In order to obtain a more accurate steady-state tracking response from the system, it is possible to modify the sliding surface to include the integral of error. In order to do so, an augmented model of the system is used. The added state variable is the integral of error of the original system.

In order to make this approach clear, the second-order system in (2.25) is reconsidered. The sliding surface for the system is refined as follows:

$$s(t) = \left(\frac{d^2}{dt^2} + k_1 \frac{d}{dt} + k_2 \right) \int_0^t e(\tau)d\tau, \tag{2.38}$$

where k_1 and k_2 are the design parameters that are to be selected as positive to provide a stable sliding surface. It is further possible to rewrite the sliding surface of (2.38) as follows:

$$s(t) = \dot{e}(t) + k_1 e(t) + k_2 \int_0^t e(\tau) d\tau. \tag{2.39}$$

Hence, as the sliding surface considered for a second-order system includes proportional, integral, and derivative parts, it is called a PID sliding surface. The augmented model for the system is considered as follows:

$$\begin{cases} \dot{x}_0 = x_1 \\ \dot{x}_1 = x_2 \\ \dot{x}_2 = f(x) + g(x)u, \end{cases} \tag{2.40}$$

where x_0 is the newly added state variable that is the integral of first state of the system.

The time derivative of the sliding surface in this case is obtained as follows:

$$\dot{s} = \ddot{r} - f(x) - g(x)u + k_1 \dot{e} + k_2 e. \tag{2.41}$$

The equivalent control signal, u_n, is defined as follows:

$$u_n = \frac{1}{g_n(x)} \left(f_n(x) - \ddot{r} - k_1 \dot{e} - k_2 e \right). \tag{2.42}$$

It is possible to rewrite (2.42) as follows:

$$- f_n(x) + \ddot{r} + k_1 \dot{e} + k_2 e + g_n u_n = 0. \tag{2.43}$$

By subtracting (2.43) from (2.41), the following equation is obtained:

$$\dot{s} = -f(x) + f_n(x) - g(x)u + g_n(x)u_n. \tag{2.44}$$

Considering the fact that $g(x) = g_n(x) + \Delta g(x)$ and $u = u_n + u_r$, (2.44) can be rewritten as follows:

$$\dot{s} = -f(x) + f_n(x) - \Delta g(x)u_n - g(x)u_r. \tag{2.45}$$

The candidate Lyapunov function is considered as follows:

$$V(s) = \frac{1}{2} s^2. \tag{2.46}$$

The time derivative of the Lyapunov function is obtained as follows:

$$\dot{V}(s) = s\dot{s} = s\left(- f(x) + f_n(x) - \Delta g(x)u_n - g(x)u_r. \right. \tag{2.47}$$
$$\leq F|s| + G|u_n| - g(x)u_r \tag{2.48}$$

The finite-time convergence property for the sliding surface requires that $s\dot{s} \leq -\eta|s|$; in order to satisfy this inequality, u_r is taken as follows:

$$u_r = \frac{1}{g_{min}} K sgn(s), \tag{2.49}$$

where the gain of the u_r signal, which is used to guarantee the robustness of system, is chosen as follows:

$$F + G + \eta \leq K. \tag{2.50}$$

This concludes the stability analysis of the sliding-mode controller with the PID sliding surface.

2.4   SMC for Higher-Order Nonlinear Systems

Consider a nonlinear dynamic system as follows:

$$\begin{cases} \dot{x}_1 & = x_2 \\ \quad \vdots \\ \dot{x}_{n-1} = x_n \\ \dot{x}_n & = f(x) + g(x)u. \end{cases}$$

The sliding manifold for such a system is defined as follows:

$$s = e^{(n-1)} + K_{n-2}e^{(n-2)} + \cdots + K_1\dot{e} + K_0 e, \tag{2.51}$$

where $K_0, K_1, \ldots, K_{n-2}$ are selected such that the roots of polynomial $\mu^{n-1} + K_{n-2}\mu^{n-2} + \cdots + K_1\mu + K_0$ have negative real values. The following example illustrates the design process of the sliding-mode controller for a nonlinear dynamic system of the order n.

Example 2.2 (*Simulation of the proposed method on a flexible joint robot*) In this section, the proposed method is simulated on a flexible joint robot. Define $\mathbf{q} = \{q_1 \dot{q}_1 q_2 \dot{q}_2\}$ as the set of generalized coordinates for the system [8], where

- $q_2 = -(1/m)\theta_1$ is the angular displacement of the rotor and m_1 is the gear ratio,
- q_1 is the angle of link, and $q_1 - q_2$ is the elastic displacement of the link.

According to the Euler–Lagrange equations, the analytical model of the flexible joint robot is derived as follows [8]:

$$\begin{cases} I\ddot{q}_1 + MgLsin\,(q_1) + k(q_1 - q_2) = 0 \\ \qquad\qquad J\ddot{q}_2 - k(q_1 - q_2) = u. \end{cases} \tag{2.52}$$

The state-space representation of the system is achieved as follows:

$$\dot{x}_1 = x_2$$
$$\dot{x}_2 = -\frac{MgL}{I}sin(x_1) - \frac{K}{I}(x_1 - x_3)$$
$$\dot{x}_3 = x_4$$
$$\dot{x}_4 = \frac{K}{J}(x_1 - x_3) + \frac{1}{J}u. \tag{2.53}$$

As can be seen from the dynamic model of the flexible joint robot, its dynamic model is not in the normal form of (2.51). However, it is possible to use input–output feedback linearization to obtain the dynamics of the system in normal form [9]. In order to do so, we need to introduce the concept of the Lie derivative. The Lie derivative [10] of a function $h(x)$ with respect to the function $f(x)$ is defined as follows:

$$L_f h(x) = \frac{\partial h}{\partial x}(x)f(x).$$

Moreover, the higher-order Lie derivative of the function $h(x)$ with respect to $f(x)$ is defined as follows:

$$L_f^k h(x) = \frac{\partial}{\partial x}(L_f^{k-1}h(x))f(x).$$

However, it is also possible that the higher-order Lie derivative is taken with respect to another function. In this case, we have the following definition:

$$L_g L_f^k h(x) = \frac{\partial}{\partial x}(L_f^k h(x))g(x).$$

Consider the nonlinear dynamic model in a general form, which is not in the normal form.

$$\dot{x} = \alpha(x) + \beta(x)u$$
$$y = h(x), \tag{2.54}$$

where $x = [x_1, x_2, \ldots, x_n]^T \in R^n$ is the state vector of the system, $\alpha(x), \beta(x) : R^n \to R^n$ are nonlinear functions that exist in the dynamic of system, and $h(x) : R^n \to R$ is the nonlinear output function of the system. The time derivative of the output y is obtained as follows:

$$\dot{y} = \frac{\partial h}{\partial x}\dot{x} = L_\alpha h(x) + L_\beta h(x)u. \tag{2.55}$$

It is assumed that $L_\beta h(x) = 0$; hence, we have $\dot{y} = L_\alpha h(x)$. The second-order time derivative of y is obtained as follows:

$$\ddot{y} = L_\alpha^2 h(x) + L_\beta L_\alpha h(x). \tag{2.56}$$

A nonlinear dynamic system is of relative degree equal to r, if the following equations hold:

$$L_\beta L_\alpha^i h(x) = 0, \ \ i = 0, 1, \ldots, r-2$$
$$L_\beta L_\alpha^{r-1} h(x) \neq 0. \tag{2.57}$$

Here, it is assumed that the relative degree is equal to n and the system does not have any zero dynamics. In this case, the nonlinear dynamic model of the system in normal form is obtained as follows:

$$\dot{x}_1 = x_2$$
$$\dot{x}_2 = x_3$$
$$\vdots$$
$$\dot{x}_n = L_\alpha^n h(x) + L_\beta L_\alpha^{r-1} h(x) u. \tag{2.58}$$

Using a similar procedure, the state space form of (2.53) in the normal form is obtained as follows:

$$\dot{z}_1 = z_2$$
$$\dot{z}_2 = z_3$$
$$\dot{z}_3 = z_4$$
$$\dot{z}_4 = -\left(\frac{MgL}{I}cosz_1 + \frac{K}{I} + \frac{K}{J}\right)z_3 + \frac{MgL}{I}\left(z_2^2 - \frac{K}{J}\right)sin(z_1) + \frac{K}{IJ}u, \tag{2.59}$$

where the parameters z_1, z_2, z_3, and z_4 are defined as follows:

$$z_1 = x_1$$
$$z_2 = x_2$$
$$z_3 = -\frac{MgL}{I}sin(x_1) - \frac{K}{I}(x_1 - x_3) \tag{2.60}$$
$$z_4 = -\frac{MgL}{I}x_2cos(x_1) - \frac{K}{I}(x_2 - x_4) .$$

The numerical values of the parameters considered in the simulation studies are $g = 9.8\,\mathrm{m/s^2}$, $M=1\,\mathrm{Kg}$, $K=1\,\mathrm{N/m}$, $J = 1\,\mathrm{Kgm^2}$, $I = 1\,\mathrm{Kgm^2}$, and $L = 1\,\mathrm{m}$. The sliding surface for the system is considered as follows:

$$s = \frac{d^3}{dt^3}e + 11\frac{d^2}{dt^2}e + 38\frac{d}{dt}e + 40e, \tag{2.61}$$

where $e = z_1 - r$, and the SMC signal is obtained as follows:

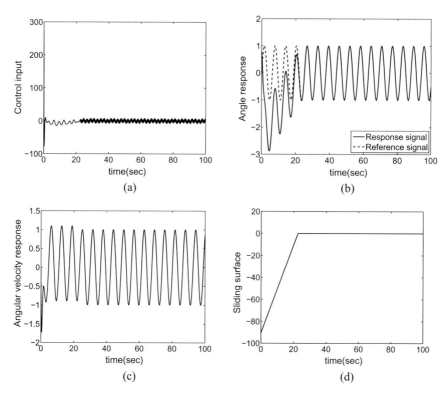

Fig. 2.5 a The control signal **b** The tracking response of the system **c** The sliding surface. **d** The trajectory of system

$$u(t) = \frac{IJ}{K} \left(\left(\frac{MgL}{I} cos(z_1) + \frac{K}{I} + \frac{K}{J} \right) z_3 - \frac{MgL}{I} \left(z_2^2 - \frac{K}{J} \right) sin(z_1) \right.$$

$$\left. - 11 \left(z_4 - \frac{d^3 r}{dt^3} \right) - 38(z_3 - \ddot{r}) - 40(z_2 - \dot{r}) - K_K sign(s) \right). \quad (2.62)$$

The simulation results are depicted in Fig. 2.5. As can be seen from the figure, the sliding surface converges to zero in 40 s. However, it is possible to decrease the finite-time convergence of the sliding manifold to zero by using a larger value for K_K. In order to do so, this parameter is multiplied by *two*, and the results are illustrated in Fig. 2.6. As can be seen from this figure, the convergence time is divided by *two* when K_K is multiplied by *two*.

Fig. 2.6 Finite time
reaching with a different K

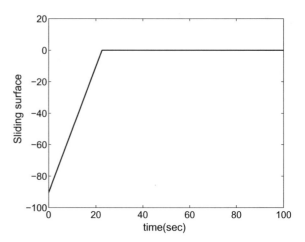

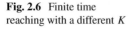

Fig. 2.6 Finite time
reaching with a different K

2.5   Adaptive Sliding-Mode Approaches

Considering the parameter variations, nonlinearity in actuators, changes in operating
points, aging, faults in the system, and time-varying disturbances in a system that
are to be controlled, an adaptive controller design is preferable, whose parameters
are tuned wisely to compensate for the aforementioned variations. Different methods
are proposed to design a classical sliding-mode controller. The two frequently used
ones are discussed in this section. In the first approach, the system parameters are
considered to be partially known and parameters of the controller are tuned to compensate for the uncertainties of the system. In the second approach, the parameters of
the system are considered to be unknown and estimated throughout the adaptation.

2.5.1   Adaptive Tuning of the Controller Parameters

As was mentioned earlier, chattering is one of the biggest problems of sliding-mode
controllers. Although a larger value for the parameter K may be used to make the
system more robust, a too large value for this parameter gives rise to a chattering
problem. Moreover, as can be seen from (2.50), the value of parameter K depends on
the maximum absolute value of system uncertainties. This means that the maximum
absolute value of system uncertainties must be known *a priori* and it must be ensured
sure that these uncertainties do not violate their predefined maximum values. In order
to deal with this problem, it is suggested that the parameter K be taken as an adaptive
parameter whose value can be determined online.

Consider the second-order nonlinear system of (2.1) with the sliding surface of
(2.3). The Lyapunov function considered for the system is modified as follows:

$$V = \frac{1}{2}s^2 + \frac{1}{2\gamma}(K - K^*)^2, \tag{2.63}$$

where $\gamma > 0$ is the learning rate considered for the parameter K and is a time-varying parameter. The time derivative of this Lyapunov function is obtained as follows:

$$\dot{V} = s\dot{s} + \frac{1}{\gamma}\dot{K}(K - K^*) \tag{2.64}$$

using a similar analysis as in Sect. 2.2 and using the control signal as in (2.13), we have

$$\dot{V} = s\dot{s} + \frac{1}{\gamma}\dot{K}(K - K^*) \le F|s| + su_r + \frac{1}{\gamma}\dot{K}(K - K^*). \tag{2.65}$$

The signal u_r, the part of control signal to ensure its robustness, is taken as $u_r = Ksgn(s)$. By adding and subtracting $K^*|s|$ from (2.65), the following equation is obtained:

$$\dot{V} \le F|s| - K^*|s| + (K^* - K)|s| + \frac{1}{\gamma}\dot{K}(K - K^*). \tag{2.66}$$

Considering the adaptation law of K as follows:

$$\dot{K} = \gamma|s|, \tag{2.67}$$

we have the following equation for (2.66):

$$\dot{V} \le F|s| - K^*|s|. \tag{2.68}$$

In order to guarantee the fast and finite time convergence of the sliding surface, K^* may be taken as follows:

$$F + \eta \le K^*. \tag{2.69}$$

This selection for the parameter K^* guarantees that

$$\dot{V} = s\dot{s} \le -\eta|s|, \tag{2.70}$$

which guarantees the robust stability of the sliding surface. Moreover, the parameter K is tuned automatically and is not needed to be selected a priori.

Example 2.3 The inverted pendulum in Example 2.1 is used to show how the adaptive parameter K can be used to lessen a priori knowledge needed to control this system.

The simulation results are depicted in Fig. 2.7. The system responds appropriately to the reference signal. As can be seen from 2.7d, the adaptive parameter K_K varies

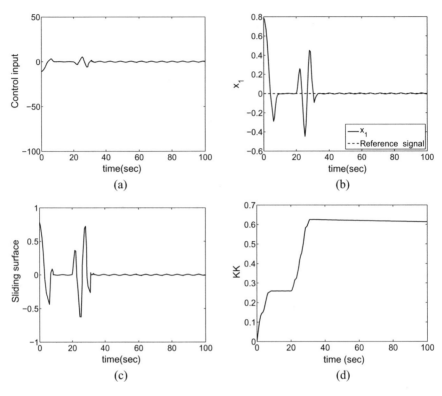

Fig. 2.7 **a** The control signal **b** The tracking response of the system **c** The sliding surface. **d** The evolution of adaptive parameter K_K

with time to find its appropriate value. It is further assumed that in the 80th second, the disturbance changes from $0.1 sin(t)$ to $0.9 sin(t)$. As can be seen from the figure, the parameter K_K varies to compensate this change in the disturbance of the system. The fact that disturbances of the system do not necessarily need to be known is the main benefit of the usage of adaptive parameter K_K.

2.5.2   Online Identification of System Parameters

We may have partial knowledge about the system plant model but the system parameter values are unknown. In this case, to calculate the equivalent control signal u_n, the parameters of the system need to be estimated. There exist two approaches: robust approach and the adaptive control approach. In the case of the robust approach, the unknown part of the system is considered as its uncertainty. The controller designed in this case may be conservative, which may result in unnecessarily large control signals. On the other hand, there exists an alternative approach, according to which, the unknown part of the system is identified and is used to compute the equivalent control signal. The following two examples illustrate the second approach.

Example 2.4 Consider the dynamic equation that governs the Duffing oscillator as follows:

$$\ddot{x} + a_0 x + a_1 x^3 + a_2 \dot{x} + f_0 cos(\omega t) = u(t),\qquad(2.71)$$

where a_0, a_1, and a_2 are considered to be unknown real parameters. In order to construct the equivalent SMC signal, the values of these parameters must be known. The sliding surface considered for the system is as follows:

$$s = \dot{e} + \lambda e.\qquad(2.72)$$

The Lyapunov function considered for the system includes the sliding mode as well as estimation errors of parameter as follows:

$$V = \frac{1}{2}s^2 + \frac{1}{2\gamma_0}\left(\hat{a}_0 - a_0\right)^2 + \frac{1}{2\gamma_1}\left(\hat{a}_1 - a_1\right)^2 + \frac{1}{2\gamma_2}\left(\hat{a}_2 - a_2\right)^2,\qquad(2.73)$$

where $\hat{a}_0$, $\hat{a}_1$, and $\hat{a}_2$ are considered to be the estimated values for a_0, a_1, and a_2, respectively. The time derivative of the Lyapunov function to derive the adaptation laws for the unknown parameters of the system to guarantee system stability is as follows:

$$\dot{V} = s\dot{s} + \frac{1}{\gamma_0}\dot{\hat{a}}_0\left(\hat{a}_0 - a_0\right) + \frac{1}{\gamma_1}\dot{\hat{a}}_1\left(\hat{a}_1 - a_1\right) + \frac{1}{\gamma_2}\dot{\hat{a}}_2\left(\hat{a}_2 - a_2\right).\qquad(2.74)$$

Considering the nonlinear dynamic of the Duffing oscillator (2.71) and the sliding surface of (2.72), we have the following equation for time derivative of the Lyapunov function:

$$\dot{V} = s\left(\ddot{r} + \hat{a}_0 x + \hat{a}_1 x^3 + \hat{a}_2 \dot{x} + u + f_0 cos(\omega t) + \lambda \dot{e}\right) + \frac{1}{\gamma_0}\dot{\hat{a}}_0\left(\hat{a}_0 - a_0\right) + \frac{1}{\gamma_1}\dot{\hat{a}}_1\left(\hat{a}_1 - a_1\right)$$
$$+ \frac{1}{\gamma_2}\dot{\hat{a}}_2\left(\hat{a}_2 - a_2\right).\qquad(2.75)$$

Since the system parameters are unknown, the equivalent control signal u_n is constructed based on the identified parameters as follows:

$$u_n(t) = -\ddot{r} - \hat{a}_0 x - \hat{a}_1 x^3 - \hat{a}_2 \dot{x} - \lambda \dot{e}.\qquad(2.76)$$

Using the equivalent control signal of (2.76), the time derivative of the Lyapunov function (2.73) is obtained as follows:

$$\dot{V} = (a_0 - \hat{a}_0)sx + (a_1 - \hat{a}_1)sx^3 + (a_2 - \hat{a}_2)s\dot{x} + f_0cos(\omega t)s + su_r + \frac{1}{\gamma_0}\dot{\hat{a}}_0(\hat{a}_0 - a_0)$$

$$+ \frac{1}{\gamma_1}\dot{\hat{a}}_1(\hat{a}_1 - a_1) + \frac{1}{\gamma_2}\dot{\hat{a}}_2(\hat{a}_2 - a_2). \tag{2.77}$$

Considering the adaptation laws of $\hat{a}_0$, $\hat{a}_1$, and $\hat{a}_2$,

$$\dot{\hat{a}}_0 = \gamma_0 sx \tag{2.78}$$

$$\dot{\hat{a}}_1 = \gamma_1 sx^3 \tag{2.79}$$

$$\dot{\hat{a}}_2 = \gamma_2 s\dot{x}, \tag{2.80}$$

we have the following equation for the time derivative of the Lyapunov function:

$$\dot{V} = su_r + sf_0cos(\omega t). \tag{2.81}$$

Considering u_r as to be equal to $u_r = -Ksgn(s)$, in which, $\eta + f_0 < K$, the following equation is obtained:

$$\dot{V} \leq -\eta|s|, \tag{2.82}$$

which guarantees the finite-time convergence of the sliding surface s to zero.

The initial conditions considered for the system are selected as to be equal to $\hat{a}_0(0) = 0.1$, $\hat{a}_1(0) = 0$, and $\hat{a}_2 = 0$. The simulation results are illustrated in Fig. 2.8. As can be seen from this figure, the parameters of the system are tuned and the state x_1 converges to its reference signal. It is to be noted that the stability analysis done for the adaptive control system guarantees the controller parameter boundedness but does not necessarily guarantee the parameter convergence to their true values. Generally, parameter convergence to their true values requires persistency excited reference and/or control signal. This may be achieved by adding artificial noise with a small amplitude to the control and reference signal.

2.5.3 Adding Robustness to the Adaptation Laws

Although the Lyapunov stability analysis guarantees the stability of the system, parameter bursting may happen, which causes internal instability in the system. This problem may happen as a result of a lack of persistency of excitation in control signal, the existence of disturbances, unmodeled dynamics, and high-frequency control signal which may excite high-frequency response of the system [6].

As can be seen from (2.67), the adaptation law of the parameter K is always positive, which may make this parameter unnecessarily large. In order to avoid these problems, it is possible to add σ-modification term to (2.67) as follows:

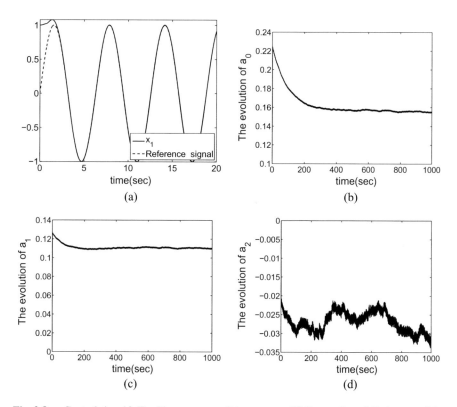

Fig. 2.8 **a** Control signal **b** Tracking response of the system **c** Sliding surface. **d** Trajectory of the system

$$\dot{K} = \gamma |s| - \nu \gamma K. \tag{2.83}$$

This modification avoids parameter-bursting by changing the parameter K in the opposite direction of its growth [5, 7]. When K has a large positive value, the term $-\nu \gamma K$ becomes a large negative value and decreases its value. When K has a large negative value, the term $-\nu \gamma K$ acts in the reverse direction and avoids instability. However, the existence of this term may force the parameter K to converge to zero if s is close to zero. In order to avoid this problem, ϵ-modification may be used to modify the adaptation law of (2.67) as follows:

$$\dot{K} = \gamma |s| - \epsilon \gamma K |s|. \tag{2.84}$$

The existence of $|s|$ in the modified adaptation law avoids the convergence of K to zero by stopping the adaptation law when s is very close to zero.

2.6 Nonlinear and Time-Varying Sliding Surfaces

In order to improve the dynamic response of the sliding-mode controller, sliding manifold can be taken as a nonlinear and/or a time-varying function of the system states. Different approaches have been proposed for the time-varying sliding surface. The first approach, which is considered in [3], suggests that instead of using a sliding surface with constant parameters, a nonlinear time-varying sliding surface can be used, which eliminates the reaching phase. In this approach, instead of making the states reach sliding manifold, a time-varying sliding manifold is designed such that system states are initially located on the sliding manifold. This avoids the high-gain control signal during the reaching mode. However, the sliding surface converges to the desired trajectory of the system in a finite time and the desired behavior of the system is achieved.

Time-varying slope for the sliding surface is considered to initially include system states. Consider the second-order nonlinear system in (2.1). The sliding mode with time-varying slope has the following form:

$$s(t) = \dot{e} + \lambda(t)e. \tag{2.85}$$

In order to ensure that $s(0) = 0$ and system states are initially located on the sliding surface, $\lambda(0)$ is selected to be equal to $-\dot{e}(0)/e(0)$. The time-varying parameter $\lambda(t)$ is designed such that after a while it converges to the desired value of the slope of the sliding surface, which defines the desired trajectory for the system.

It is further possible to define a time-varying intercept for the sliding surface. For instance, consider the following sliding surface for the second-order system of (2.1):

$$s(t) = \dot{e}(t) + \lambda e(t) - \alpha(t)$$
$$s(0) = \dot{e}(0) + \lambda e(0) - \alpha(0), \tag{2.86}$$

where λ is the fixed slope of the sliding surface and $\alpha(0)$ is selected such that it ensures that $s(0) = 0$. In this case,

$$\alpha(0) = \dot{e}(0) + \lambda e(0). \tag{2.87}$$

Furthermore, the terminal value of α is selected as to be equal to *zero*, to ensure the desired behavior of the system.

More time-varying sliding surfaces with variable slope or time-varying axis-intercept can be chosen. For instance, the following time-varying sliding surface can be chosen for a general class of nonlinear systems [1]:

$$s(t) = \left(\frac{d}{dt} + \lambda\right)^{n-1} e(t) - \frac{2}{\pi}\left(\frac{\pi}{2} - tan^{-1}(t)\right)\left(\frac{d}{dt} + \lambda\right)^{n-1} e(0). \tag{2.88}$$

This time-varying sliding surface ensures that $s(0) = 0$, which means that the system states are initially on the sliding surface.

$$\lim_{t \to \infty} \left(\frac{\pi}{2} - tan^{-1}(t) \right) = 0 \tag{2.89}$$

which guarantees the convergence of sliding surface to zero after sufficiently large time.

Another possible time-varying sliding surface is as follows:

$$E(t) = e(t)(1 - \eta(t),) \tag{2.90}$$

where $E(t)$ is an innovative error and $\eta(t)$ is initially equal to *one* and finally converges to zero. For example, $\eta(t)$ can be chosen as follows:

$$\eta(t) = \gamma exp(z(t)), \tag{2.91}$$

where

$$\gamma(t) = \sum_{j=0}^{n-1} \frac{1}{j!} \gamma^{(j)}(t - t_0)$$
$$z(t) = -\beta t, \tag{2.92}$$

where $\gamma^{(j)}$ is defined as follows:

$$\gamma^{(j)} = \frac{d^j}{dt^j} \gamma. \tag{2.93}$$

The sliding surface for the system is considered to be of the following form:

$$s(t) = \left(\frac{d}{dt} + \lambda \right)^{n-1} E(t), \quad \lambda > 0. \tag{2.94}$$

2.7 Terminal SMC

The concept of terminal SMC was proposed in [11]. Consider the following differential equation:

$$s(t) = \dot{x}_1(t) + \beta x_1^{p/q}(t) = 0, \quad \beta > 0, \ p > q > 0. \tag{2.95}$$

The solution to this nonlinear differential equation is as follows:

$$x_1(t) = \left(-\frac{\beta(p-q)}{p}t + x_1^{1-\frac{q}{p}}(0) \right)^{\frac{p}{p-q}}. \tag{2.96}$$

It can be easily shown that this signal converges to zero in finite time t_s which is equal to $t_s = \frac{p}{\beta(p-q)}|x_1(0)|^{1-\frac{q}{p}}$. Consider the nth-order nonlinear system in the following form:

$$\frac{d^n x(t)}{dt^n} = f(x(t)) + g(x(t))u(t), \tag{2.97}$$

where $f(x(t)) : R^n \to R$ $g(x(t)) : R^n \to R$ are two nonlinear functions. The terminal sliding surface is chosen sequentially as follows:

$$s_1 = \dot{s}_0 + \beta_0 s_0^{q_0/p_0}$$
$$s_2 = \dot{s}_1 + \beta_1 s_1^{q_1/p_1}$$
$$\vdots$$
$$s_{n-1} = \dot{s}_{n-2} + \beta_{n-2} s_{n-2}^{q_{n-2}/p_{n-2}}, \tag{2.98}$$

where $s_0 = x_1 - x_d$. The control signal is chosen such that $s_{n-1}\dot{s}_{n-1} \leq -\eta|s_{n-1}|$, which guarantees the convergence of s_{n-1} to zero in finite time. The time it takes for s_{n-1} to converge to zero is equal to $s_{n-1}(0)/\eta$. Once s_{n-1} is reached, the parameters $s_{n-2}, s_{n-3}, \ldots, s_0$ reach zero in finite time. The time it takes for the surface s_{n-2} to converge to zero is obtained as follows:

$$t_{s-2} = \frac{P_{n-1}}{\beta(p_{n-1} - q_{n-1})}|s_{n-2}(s_{n-1}(0)/\eta)|^{1-\frac{q_{n-1}}{p_{n-1}}}. \tag{2.99}$$

The sliding surface s_{n-3} is the next surface to reach zero with the reaching time equal to

$$t_{s-3} = \frac{P_{n-2}}{\beta(p_{n-2} - q_{n-2})}|s_{n-2}(t_{s-2})|^{1-\frac{q_{n-2}}{p_{n-2}}}. \tag{2.100}$$

Subsequently, the reaching time for the other sliding surfaces is obtained as follows:

$$t_{s-i} = \frac{Pn - i + 1}{\beta(p_{n-i+1} - q_{n-i+1})}|s_{n-i+1}(t_{n-i+1})|^{1-\frac{q_{n-i+1}}{p_{n-i+1}}}, \quad i = 2, \ldots, n \tag{2.101}$$

with t_{s-1} equal to $s_{n-1}(0)/\eta$. The total reaching time to the equilibrium is obtained as follows:

$$T = \sum_{i=1}^{n} t_{s-i}. \tag{2.102}$$

The control signal $u(t)$ is considered to be composed of two terms as $u(t) = u_n(t) + u_r(t)$. The following equation for the time derivative of s_{n-1} is obtained:

$$\dot{s}_{n-1} = \dot{x}_n + \sum_{k=0}^{n-2} \beta_k \frac{d^{n-k-1}}{dt^{n-k-1}} s_k^{q_k/p_k} - x_d^{(n)}. \tag{2.103}$$

By applying (2.97)–(2.103), the following equation is obtained:

$$\dot{s}_{n-1} = f(x) + g(x)u + \sum_{k=0}^{n-2} \beta_k \frac{d^{n-k-1}}{dt^{n-k-1}} s_k^{q_k/p_k} - x_d^{(n)}. \tag{2.104}$$

The equivalent control signal is as follows:

$$u_n = \frac{1}{g(x)} \left(x_d^{(n)} - f(x) - \sum_{k=0}^{n-2} \beta_k \frac{d^{n-k-1}}{dt^{n-k-1}} s_k^{q_k/p_k} \right). \tag{2.105}$$

As was mentioned earlier, in order to obtain the finite time convergence of s_{n-1} to zero, the control signal u_r is taken such that $\dot{s}_{n-1}s_{n-1} \leq -\eta|s_{n-1}|$. Hence, u_r is taken as to be equal to $-K sgn(s_{n-1})$ where K is selected as $K > \eta$.

It is possible that the term

$$\frac{d^{n-k-1}}{dt^{n-k-1}} s_k^{q_k/p_k} \tag{2.106}$$

causes singularity in the equivalent control signal. In order to avoid this problem in control signal, p_k and q_k must satisfy the following inequality:

$$\frac{q_k}{p_k} > \frac{n-k-1}{n-k}. \tag{2.107}$$

It is also possible to consider uncertainties in $f(x)$ and $g(x)$, which requires that the control signal u_r be modified to control the system under uncertain conditions.

2.8 SMC with Mismatched Uncertainties

Up till now, the uncertainties studied in the book were matched uncertainties. However, it is possible that the uncertainties of the system are mismatched and the control signal cannot directly deal with such uncertainties. It is possible to use backstepping to extend use of the SMC systems to a more general class of nonlinear systems [4]. Consider the nth-order nonlinear dynamic system with mismatched uncertainties as follows:

$$\begin{cases} \dot{x}_1 = x_2 + \delta_1(x_1) + f_1(x_1) \\ \dot{x}_2 = x_3 + \delta_2(x_1, x_2) + f_2(x_1, x_2) \\ \quad\vdots \\ \dot{x}_n = u + \delta_n(x_1, x_2, \ldots, x_{n-1}) + f_n(x_1, x_2, \ldots, x_{n-1}), \end{cases} \quad (2.108)$$

where $f_1(x_1)$, $f_2(x_1, x_2)$, $\ldots$, $f_n(x_1, x_2, \ldots, x_{n-1})$ are nonlinear known functions and $\delta_1(x_1)$, $\delta_2(x_1, x_2)$, $\ldots$, $\delta_n(x_1, x_2, \ldots, x_{n-1})$ are mismatched uncertainties considered for the system, which satisfy the following equations:

$$0 < |\delta_1(x_1)| \leq \gamma_1(x_1)$$
$$0 < |\delta_2(x_1, x_2)| \leq \gamma_2(x_1, x_2)$$
$$\vdots$$
$$0 < |\delta_n(x_1, x_2, \ldots, x_{n-1})| \leq \gamma_n(x_1, x_2, \ldots, x_{n-1}), \quad (2.109)$$

where $\gamma_1(x_1)$, $\gamma_2(x_1, x_2)$, $\ldots$, $\gamma_n(x_1, x_2, \ldots, x_{n-1})$ are the known upper bounds of mismatched system uncertainties.

In the first step, a Lyapunov function is introduced to guarantee the finite-time convergence of x_1 to *zero* as follows:

$$V_1 = \frac{1}{2}x_1^2. \quad (2.110)$$

The time derivative of this Lyapunov function is obtained as follows:

$$\dot{V}_1 = x_1\dot{x}_1 = x_1(x_2 + \delta_1(x_1) + f_1(x_1)). \quad (2.111)$$

Let x_2 be the control signal to provide the finite time stability of x_1. This signal is considered as follows:

$$x_2 = -\gamma_1(x_1)sgn(x_1) - f_1(x_1) = \phi_1(x_1). \quad (2.112)$$

Using (2.112), we have the following equation for (2.111):

$$\dot{V}_1 \leq -\gamma_1(x_1)|x_1|. \quad (2.113)$$

However, the state variable x_2 cannot be directly assigned a function and the algorithm must be continued to obtain an appropriate control signal for the system in which case (2.111) is modified as follows:

$$\dot{V}_1 = x_1(x_2 + \delta_1(x_1) + f_1(x_1) - \phi_1(x_1) + \phi_1(x_1)) \leq -\gamma_1(x_1)|x_1| + (x_2 - \phi_1(x_1))x_1. \quad (2.114)$$

In the second step, the second Lyapunov function is considered as follows:

$$V_2 = V_1 + \frac{1}{2}(x_2 - \phi_1(x_1))^2. \tag{2.115}$$

The second term in the Lyapunov function is considered to guarantee the convergence of state variable x_2 to $-K_1 x_1 - \gamma_1(x_1)sgn(x_1) - f_1(x_1)$ which guarantees the finite time convergence of the state variable x_1 to $zero$. Equation (2.115) can be rewritten as follows:

$$V_2 = \frac{1}{2}(x_1 - x_2 + x_2)^2 + \frac{1}{2}(x_2 - \phi_1(x_1))^2. \tag{2.116}$$

The time derivative of the Lyapunov function V_2 is obtained as follows:

$$\dot{V}_2 = (x_1 - \phi_1(x_1))\dot{x}_1 + \phi_1(x_1)\dot{x}_1 + (x_2 - \phi_1(x_1))(\dot{x}_2 - \dot{\phi}_1(x_1)). \tag{2.117}$$

Considering (2.108) and (2.113), the time derivative of the Lyapunov function V_2 can be rewritten as follows:

$$\dot{V}_2 = -\gamma(x_1)|x_1| + (x_2 - \phi_1(x_1))(x_1 + x_3 + \delta_2(x_1, x_2) + f_2(x_1, x_2) - \dot{\phi}_1(x_1)). \tag{2.118}$$

In order to make the time derivative of V_2 negative, it is possible to take x_3 as follows:

$$x_3 = \phi_2(x_1, x_2) = -\gamma_2(x_1, x_2)sgn(x_2 - \phi_1(x_1)) - f_2(x_1, x_2) + \dot{\phi}_1(x_1). \tag{2.119}$$

Considering this equation, the time derivative of the Lyapunov function V_2 is obtained as follows:

$$\dot{V}_2 = -\gamma(x_1)|x_1| - \gamma_2(x_1, x_2)|x_2 - \phi_1(x_1)|. \tag{2.120}$$

The desired values for the other state variables of the system is obtained as follows:

$$\begin{aligned} x_{i+1} &= \phi_i(x_1, x_2, \ldots, x_i) \\ &= -\gamma_i(x_1, x_2, \ldots, x_i)sgn(x_i - \phi_{i-1}(x_1, x_2, \ldots, x_{i-1})) \\ &\quad - f_i(x_1, x_2, \ldots, x_{i-1}) + \dot{\phi}_{i-1}(x_1, x_2, \ldots, x_{i-1}), \end{aligned} \tag{2.121}$$

where $\phi_0 = 0$, and the control signal to guarantee the robust stability of the system is obtained as follows:

$$\begin{aligned} u &= \phi_n(x_1, x_2, \ldots, x_n) \\ &= -\gamma_n(x_1, x_2, \ldots, x_n)sgn(x_n - \phi_{n-1}(x_1, x_2, \ldots, x_{n-1})) \\ &\quad - f_n(x_1, x_2, \ldots, x_{n-1}) + \dot{\phi}_{n-1}(x_1, x_2, \ldots, x_{n-1}). \end{aligned} \tag{2.122}$$

This control signal guarantees that the time derivative of the following Lyapunov function is negative.

$$V = \frac{1}{2}x_1^2 + \frac{1}{2}\sum_{i=2}^{n}(x_i - \phi_{i-1}(x_1, x_2, \ldots, x_{i-1}))^2. \tag{2.123}$$

The time derivative of the Lyapunov function V is obtained as follows:

$$\dot{V} = -\gamma_1(x_1)|x_1| - \sum_{i=2}^{n}\gamma_i(x_1, x_2, \ldots, x_{i-1})|x_i - \phi_{i-1}(x_1, x_2, \ldots, x_{i-1})|. \tag{2.124}$$

The following analysis illustrates the sliding motion caused by the control signal in the system. In this analysis, it is assumed that δ_is, the uncertain parts of the dynamics of system (2.108), are equal to *zero*. We have the following equation for the time derivative of x_1:

$$\dot{x}_1 = x_2 - \phi_1(x_1) + f_1(x_1) + \phi_1(x_1), \tag{2.125}$$

where $\phi_1(x_1) = -f_1(x_1) - K_1 sgn(x_1)$. It is possible to substitute $x_2 - \phi_1(x_1)$ with a new variable z_2. Using this change in variable, the dynamics of x_1 is obtained as follows:

$$\dot{x}_1 = z_2 - K_1 sgn(x_1). \tag{2.126}$$

The dynamics of z_2 is obtained as follows:

$$\begin{aligned}
\dot{z}_2 &= \dot{x}_2 - \frac{\partial \phi_1(x_1)}{\partial x_1}\dot{x}_1 \\
&= x_3 + f_2(x_1, x_2) - \frac{\partial \phi_1(x_1)}{\partial x_1}(x_2 + f_1(x_1)).
\end{aligned} \tag{2.127}$$

Let $\phi_2(x_1, x_2)$ be taken as follows:

$$\phi_2(x_1, x_2) = -f_2(x_1, x_2) + \frac{\partial \phi_1(x_1)}{\partial x_1}(x_2 + f_1(x_1)) - K_2 sgn(z_2), \tag{2.128}$$

the following equation is then obtained:

$$\dot{z}_2 = x_3 - \phi_2(x_1, x_2) - K_2 sgn(z_2). \tag{2.129}$$

It is possible to make a new change in variables as $z_3 = x_3 - \phi_2(x_1, x_2)$ to obtain the following equation:

$$\dot{z}_2 = z_3 - K_2 sgn(z_2). \tag{2.130}$$

This process can be followed for all other differential equations to obtain the following dynamic equations for the system:

$$\dot{x}_1 = z_2 - K_1 sgn(x_1)$$
$$\dot{z}_2 = z_3 - K_2 sgn(z_2)$$
$$\vdots$$
$$\dot{z}_{n-1} = z_n - K_{n-1} sgn(z_{n-1})$$
$$\dot{z}_n = u + f_n(x_1, x_2, \ldots, x_n) - \sum_{i=1}^{n-1} \frac{\partial \phi_{n-1}(x_1, \ldots, x_{n-1})}{\partial x_i}(x_{i+1} + f_i(x_1, \ldots, x_i)). \quad (2.131)$$

For the finite-time convergence of z_n to $zero$, the control signal u is taken as follows:

$$u = -f_n(x_1, x_2, \ldots, x_n) + \sum_{i=1}^{n-1} \frac{\partial \phi_{n-1}(x_1, \ldots, x_{n-1})}{\partial x_i}(x_{i+1} + f_i(x_1, \ldots, x_i)) - K_n sgn(z_n). \quad (2.132)$$

This control signal guarantees that $z_n \dot{z}_n < -K_n|z_n|$, which means finite-time convergence of z_n to $zero$. As soon as z_n converges to zero, the dynamic equation of z_{n-1} becomes $\dot{z}_{n-1} = -K_{n-1} sgn(z_{n-1})$, which similarly guarantees the finite-time convergence of z_{n-1} to $zero$. Similarly, all other states of the system converge to $zero$ in finite time.

Although the $sign$ function that is used in the design of this type of controller guarantees finite-time convergence of system parameters, it cannot be used directly in these equations. The function ϕ needs to be sufficiently differentiable. In order to fulfill these conditions, the following hyperbolic tangent function is used:

$$sig(x) = \frac{exp(x) - exp(-x)}{exp(x) + exp(-x)}. \quad (2.133)$$

This function approximates the $sign$ function, and its time derivative of any order is smooth.

The following example illustrates the process of a backstepping design for sliding-mode controller.

Example 2.5 The dynamic equations governing the Lorenz system are as follows:

$$\dot{x}_1 = a_1(x_2 - x_1)$$
$$\dot{x}_2 = x_1(a_2 - x_3) - x_2$$
$$\dot{x}_3 = x_1 x_2 - a_3 x_3 + u. \quad (2.134)$$

The following Lyapunov function is considered in the first step:

$$V_1 = \frac{1}{2}e^2. \quad (2.135)$$

The time derivative of the Lyapunov function V_1 is obtained as follows:

$$\dot{V}_1 = e\dot{e} = e(\dot{x}_1 - \dot{r}) = e(a_1 x_2 - a_1 x_1 - \dot{r}). \quad (2.136)$$

Considering the dynamic equation of the Lorenz system as in (2.134), the following equation for the time derivative of error is obtained:

$$\dot{V}_1 = e(a_1 x_2 - a_1 x_1 - \dot{r} + a_1 \phi_1 - a_1 \phi_1), \tag{2.137}$$

where ϕ_1 is considered to be as

$$\phi_1 = x_1 - \frac{1}{a_1}\dot{r} - K_1 sig(\eta_1 e). \tag{2.138}$$

By applying ϕ_1 as in (2.138)–(2.137), the following equation is obtained:

$$\dot{V}_1 = -K_1 e.sig(\eta_1 e) + a_1 e(x_2 - \phi_1). \tag{2.139}$$

A change in the parameters is defined as $z_2 = x_2 - \phi_1$. Using this change in parameters, we obtain the following equation:

$$\dot{V}_1 = -K_1 e.sig(\eta_1 e) + a_1 e z_2. \tag{2.140}$$

It is now possible to take the second step by defining the Lyapunov function V_2 as follows:

$$V_2 = V_1 + \frac{1}{2}z_2^2. \tag{2.141}$$

Using the dynamic equation of the Lorenz system as in (2.134), the time derivative of the variable z_2 is obtained as follows:

$$\dot{z}_2 = \dot{x}_2 - \dot{\phi}_1 = a_2 x_1 - x_1 x_3 - x_2 - a_1 x_2 + a_1 x_1 + \frac{1}{a}\ddot{r} + K_1 \eta_1 \dot{e}(1 - sig^2(\eta_1 e)). \tag{2.142}$$

The time derivative of the Lyapunov function V_2 is obtained as follows:

$$\dot{V}_2 = \dot{V}_1 + z_2 \dot{z}_2. \tag{2.143}$$

Considering (2.142), the time derivative of the Lyapunov function V_2 can be rewritten as follows:

$$\begin{aligned} \dot{V}_2 = & -K_1 e.sig(\eta_1.e) \\ & + z_2 \Big(a_1 e + a_2 x_1 - x_1 x_3 - x_2 - a_1 x_2 + a_1 x_1 \\ & + \frac{1}{a}\ddot{r} + K_1 \eta_1 \dot{e}(1 - sig^2(\eta_1 e)) + \phi_2 - \phi_2 \Big), \end{aligned} \tag{2.144}$$

where the parameter ϕ_2 is defined as follows:

$$\phi_2 = -a_1 e - a_2 x_1 + x_2 + a_1 x_2 - a_1 x_1$$
$$- \frac{1}{a}\ddot{r} - K_1 \eta_1 \dot{e}\big(1 - sig^2(\eta_1 e)\big) - K_2 . sig(\eta_2 z_2). \tag{2.145}$$

Substituting (2.145) into (2.144), the following equation is obtained:

$$\dot{V}_2 = -K_1 e . sig(\eta_1 . e) - K_2 z_2 . sig(\eta_2 z_2) + z_2 z_3. \tag{2.146}$$

A new change in variables is considered as $z_3 = -x_1 x_3 - \phi_2$, whose time derivative of error is obtained as follows:

$$\dot{z}_3 = -\dot{x}_1 x_3 - x_1 \dot{x}_3 - \frac{\partial \phi_2}{\partial x_1}\dot{x}_1 - \frac{\partial \phi_2}{\partial x_2}\dot{x}_2 - \frac{1}{a}r^{(3)} - \frac{\partial \phi_2}{\partial z_2}\dot{z}_2 - \frac{\partial \phi_2}{\partial e}\dot{e} - \frac{\partial \phi_2}{\partial \dot{e}}\ddot{e}. \tag{2.147}$$

At the final step, the Lyapunov function is considered to be as follows:

$$V = V_2 + \frac{1}{2}z_3^2. \tag{2.148}$$

The time derivative of the Lyapunov function V is obtained as follows:

$$\dot{V} = \dot{V}_2 + z_3 \dot{z}_3. \tag{2.149}$$

Considering (2.147), the following equation for the time derivative of the Lyapunov function V is obtained as follows:

$$\dot{V} = -K_1 e . sig(\eta_1 e) - K_2 z_2 . sig(\eta_2 z_2) + z_2 z_3$$
$$+ z_3\left(-\dot{x}_1 x_3 - x_1 \dot{x}_3 - \frac{\partial \phi_2}{\partial x_1}\dot{x}_1 - \frac{\partial \phi_2}{\partial x_2}\dot{x}_2 - \frac{1}{a}r^{(3)} - \frac{\partial \phi_2}{\partial z_2}\dot{z}_2 - \frac{\partial \phi_2}{\partial e}\dot{e} - \frac{\partial \phi_2}{\partial \dot{e}}\ddot{e}\right). \tag{2.150}$$

Considering the dynamic equation of the Lorenz system as in (2.134), the following equation is obtained:

$$\dot{V} = -K_1 e . sig(\eta_1 e) - K_2 z_2 . sig(\eta_2 z_2) + z_2 z_3$$
$$+ z_3\left(-\big(a_1(x_2 - x_1)\big)\big)x_3 - x_1\big(x_1 x_2 - a_3 x_3 + u\big)\right.$$
$$\left. - \frac{\partial \phi_2}{\partial x_1}\dot{x}_1 - \frac{\partial \phi_2}{\partial x_2}\dot{x}_2 - \frac{1}{a}r^{(3)} - \frac{\partial \phi_2}{\partial z_2}\dot{z}_2 - \frac{\partial \phi_2}{\partial e}\dot{e} - \frac{\partial \phi_2}{\partial \dot{e}}\ddot{e}\right). \tag{2.151}$$

In order for the time derivative of the Lyapunov function V to be negative, the following control signal is derived:

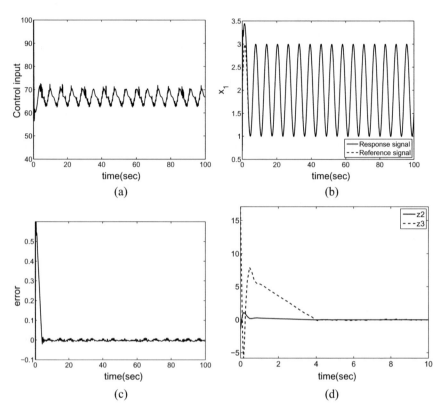

Fig. 2.9 **a** Control signal **b** Tracking response of the system **c** Sliding surface. **d** Trajectory of the system

$$u = -x_1 x_2 + a_3 x_3 + \frac{1}{x_1}\left(-a_1(x_2 - x_1)x_3 + z_2 - \frac{\partial \phi_2}{\partial x_1}\dot{x}_1 - \frac{\partial \phi_2}{\partial x_2}\dot{x}_2 \right.$$
$$\left. - \frac{1}{a}r^{(3)} - \frac{\partial \phi_2}{\partial z_2}\dot{z}_2 - \frac{\partial \phi_2}{\partial e}\dot{e} + K sign(z_3) - \frac{\partial \phi_2}{\partial \dot{e}} \right). \tag{2.152}$$

Figure 2.9 illustrates the obtained results. As can be seen from the figure, the error converges to zero in a finite time. Another simulation is done in which the reference signal for the parameter x_1 includes *zero*. Despite successful control of the system, large control signal values are observed. This is mainly because system state x_1 exists in the denumerator of the control signal and whenever it becomes too close to *zero*, a large control signal is observed. This issue must be taken into account and in the simulations, a lower bound for $|x_1|$ must be set (Fig. 2.10).

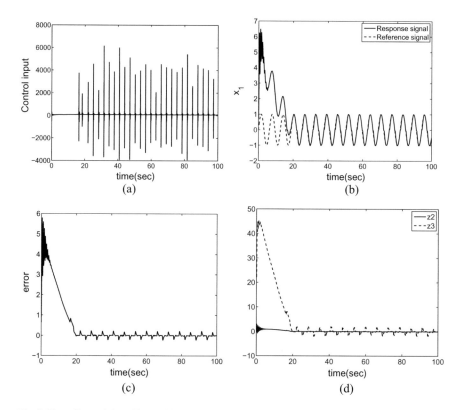

Fig. 2.10 **a** Control signal **b** Tracking response of the system **c** Sliding surface. **d** Trajectory of the system

References

1. Al-khazraji, A., Essounbouli, N., Hamzaoui, A., Nollet, F., Zaytoon, J.: Type-2 fuzzy sliding mode control without reaching phase for nonlinear system. Eng. Appl. Artif. Intell. **24**(1), 23–38 (2011)
2. Cho, Y.W., Park, C.W., Park, M.: An indirect model reference adaptive fuzzy control for SISO Takagi-Sugeno model. Fuzzy Sets Syst. **131**(2), 197–215 (2002)
3. Choi, S.B., Park, D.W., Jayasuriya, S.: A time-varying sliding surface for fast and robust tracking control of second-order uncertain systems. Automatica **30**(5), 899–904 (1994)
4. Davila, J.: Exact tracking using backstepping control design and high-order sliding modes. IEEE Trans. Autom. Control **58**(8), 2077–2081 (2013)
5. Farrell, J.A., Polycarpou, M.M.: Adaptive Approximation Based Control: Unifying Neural, Fuzzy and Traditional Adaptive Approximation Approaches, vol. 48. Wiley, Hoboken (2006)
6. Hovd, M., Bitmead, R.R.: Directional leakage and parameter drift. Int. J. Adapt. Control Signal Process. **20**(1), 27–39 (2006)
7. Ioannou, P.A., Sun, J.: Robust Adaptive Control. Courier Corporation, North Chelmsford (2012)
8. Kanoh, H., Tzafestas, S., Lee, H., Kalat, J.: Modelling and control of flexible robot arms. In: 1986 25th IEEE Conference on Decision and Control, pp. 1866–1870. IEEE (1986)
9. Khalil, H.K., Grizzle, J.: Nonlinear Systems, vol. 3. Prentice Hall, New Jersey (1996)

10. Vidyasagar, M.: Nonlinear Systems Analysis, vol. 42. SIAM, Philadelphia (2002)
11. Yu, X., Zhihong, M.: Fast terminal sliding-mode control design for nonlinear dynamical systems. IEEE Trans. Circuits Syst.[2]I: Fundam. Theory Appl. **49**(2) (2002)

Chapter 3
Fuzzy Logic Systems

3.1 Introduction

The word "fuzzy" means imprecisely defined, confused, and vague. Thus *fuzzy logic is a precise logic of imprecision and approximate reasoning* [37]. FLSs are systems to be precisely defined, and SMFC is a special kind of nonlinear control with precise formulation and well-established mathematical analysis [32].

Lotfi Zadeh proposed fuzzy logic in his paper entitled "fuzzy sets," which was published in the journal of *Information and Control* [36] as an extension to conventional logic, in which, the degree of membership of an input can be given any value from the interval of [0, 1] rather than the discrete values of *zero* and *one*. FLSs can deal with uncertainty and vagueness in real-time systems. Moreover, it has the possibility to represent human knowledge in terms of a system that can work independent of the expert. Fuzzy logic has been implemented in different applications during the last five decades.

The first implementation of fuzzy logic as a controller on a real-time system dates back to 1975 when Mamdani and Assilian implemented [17, 32] fuzzy logic to control a steam engine. Fuji electric water purification plant in 1980, fuzzy robots, and the self-parking car were the first applications of FLSs in industry [30, 32].

3.1.1 Fuzzy Logic and Control

There are different requirements for a successful control system; the most important of which is stability. There exist different definitions for stability, from which, Lyapunov stability, asymptotic stability, and exponential stability are considered [10]. Moreover, the stability of the system can be local, which means that it is only valid for system state initial conditions that are located in a certain region. On the other hand, the system is globally stable, which means that it is stable regardless of its state initial values. Stability analysis of the system may impose different conditions

© Springer Nature Switzerland AG 2021
M. Ahmadieh Khanesar et al., *Sliding-Mode Fuzzy Controllers*, Studies in Systems, Decision and Control 357, https://doi.org/10.1007/978-3-030-69182-0_3

for the parameters used in the control system. Other than the stability of the system, tracking performance is of a high importance.

A control system must make the system follow a given reference signal. The performance indexes, such as integral of squared error, settling time error, and speed-related performance indexes such as rise time, settling time, and overshoot, are among indexes that are paid the most attention.

Due to simplifications made during modeling, unmodeled dynamics, uncertainties in the parameter values, and other uncertainties in the system, the system model may mismatch the real system. Time variations in the system parameters may also cause such mismatches between the modeled and the real system. On the other hand, from implementation point of view, the controller can be implemented in an analog or digital form. Analogue electrical parts and components suffer from tolerances, which may result in a mismatch between the implemented controller and what has been designed. Digital implementation is often more accurate as precise central processing units with large number of binary digits are used. However, digital controllers work in discrete time, which results in an inevitable level of noise to exist in the control system. Quantization error imposed by digital implementation of controller may occur specially if processing units with low precision are used. Time delays due to analog-to-digital conversion and digital-to-analog conversion are another issue a digital controller may face during its implementation. Hence, the controller must be robust enough to deal with uncertainties in modeling and implementation.

The response of a system subject to noise and disturbances is another indication of how well its controller performs. There exist different sources of noise in a control system. Measurement noise and noise due to temperature variations can be mentioned as noise sources. Disturbance is basically an unwanted and unmeasurable signal with a frequency close to a frequency of the reference signal acting on the system, which may disturb or even destabilize the whole system.

James Watt invented a fly wall governor, which is the first well-known closed-loop control system, in 1788 [4, 18]. Maxwell, Routh, and Hurwitz further developed the stability analysis of differential equations in the late nineteenth century. Bode, Nyquist, and Nichols developed different methods in the frequency domain [4, 18]. Since the place of closed-loop control systems is very important and can determine the behavior of the control system both in transient response and its setting time, Evans developed the root locus method to visualize the effect of variation in the controller parameter on the closed-loop poles of the system. Frequency domain methods as well as root locus methods are still useful methods to analyze the system and design an appropriate controller for it.

Robust control approaches with fixed controller gains are a class of solutions to deal with uncertainties, unmodeled dynamics, and time-varying parameters. In such cases, the modeling process is of high importance; it is required that all possible circumstances and parameter variations be foreseen to design an appropriate fixed gain controller. Robust controllers with fixed gains are conservative controllers as they need to cover the worst cases, which may result in losing performance and require applying high controller gain.

In the presence of unknown dynamics of the system as well as slowly varying parameters in the system, it is required to have some methods to update the controller parameters accordingly. To maintain the performance of the system in the presence of plant uncertainties and disturbances, it is required to inject certain degree of adaptability to the control system. A well-known example of systems with slowly varying parameters is the airplane, which burns its fuel during flight while being exposed to different attitudes, air pressure, and densities, which totally change the plant model. Such a system is a nonlinear system with time-varying parameters. In comparison to fixed gain robust control approaches for dealing with uncertainties, since adaptive control systems have the possibility to adapt their parameters with respect to parameter variations, they can maintain higher performance. The history of adaptive control systems dates back to the 1950s when the MIT rule was one of the most popular approaches to update the parameters of a system. The stability analysis became vital for adaptive control systems as changes in the parameters could result in unstable response in the system. In the presence of high adaptation gains, the system may lose its stability if a short-term disturbance or noise acts on the system. In the late 1970s and early 1980s, stability proof of adaptive control systems appeared [1]. Since then, stability analysis and robust design became the most important part of any proposed adaptive control approach.

PID controllers are widely used in the industry because of their feature that can deal with steady state error and improvements that they may achieve in the transient response of the system. There exist different methods to tune the parameters of such controller from which Ziegler–Nichols and Chien–Hrones–Reswick formula are two commonly used algorithms [2]. The ease of implementation of such a controller, both in analogue and digital form, is another reason behind its widespread usage. However, the performance obtained using such a controller is limited. Moreover, the linear nature of a such controller makes it vulnerable to performance loss in the presence of high nonlinearities. Fuzzy logic-based tuning methods are introduced to tune the parameters of the PID controller online based on some parameters, such as the current value of error and its time derivative. The outputs of such FLSs are the gains of the controller [22]. This intelligent method to tune the parameters of the controller makes the controller a nonlinear and model-free one, which may capture more nonlinearities and maintain higher performance for the system.

FEL-type of controllers is another successful NNC and T1FLC approach which is successfully applied to different industrial systems such as antilock braking systems [13]. This method involves a classical fixed controller that is designed to guarantee the robust stability of the system and an adaptive intelligent feedforward controller in the form of a neural network or T1FLSs acting in parallel to the classical controller. While it is the responsibility of the classical controller to guarantee system stability, that of the intelligent system is to improve the performance of the system. The intelligent controller in this case learns the inverse dynamics of the plant [25]. The FEL controller is conceived of as an adaptive control technique that has been successfully applied to industrial plants. In [25, 26, 28], FEL is used to train a neural network and is simulated on an inverted pendulum. In addition, to achieve a better performance, a variable learning rate is used to tune the parameters of neural networks. Enhanced control of n-degree of freedom robotic manipulator and load frequency control in interconnected power system using FEL approaches has been considered in [6, 27], respectively. One of the most important issues related to the design of FEL controllers is to maintain the stability of the system after adding the intelligent system. The stability of FEL as applied to stable and stably invertible linear systems is proved [23]. Moreover, the stability property of FEL for a class of nonlinear dynamical systems is considered [24].

It is possible to prove the general function approximation property of FLSs using the Stone Weierstrass theorem under sufficient conditions for the function to be approximated and the FLS. Such properties include smoothness in the function to be approximated and sufficiently large number of fuzzy rules for the FLS [35]. Using such a property, it is possible to use FLSs to approximate the dynamic behavior of the system to adaptively control the system. In these approaches, it is possible to apply Lyapunov theorem-based stability analysis along with well-known nonlinear control approaches such as model reference adaptive control [14], SMC methods [16], and back-stepping [29] to control the system with high performance.

TSFLSs is a strong control structure used to model nonlinear systems. Using this model, it is possible to construct a nonlinear dynamic model for the system with locally linear systems. It is, therefore, easier to control such a system using parallel distributed control approach. The controller in the form of TSFLSs uses the same antecedent part as the model of the system [31]. It is required to use linear matrix inequality approaches to stabilize the system. Although the controller benefits from rigorous stability analysis, the approach is classified within model-based approaches. As time delays occur in networked control systems, the stability analysis of the system in the presence of time delays and uncertainties in the TSFLSs have been considered in several papers.

3.1.2 Boolean Versus Fuzzy Sets

Let U be the universe of discourse, which may be either discrete or continuous. A Boolean set is represented by the membership degrees of *zero* or *one* as follows:

$$\mu_A(x) = \begin{cases} 1 \ if \ x \in A \\ 0 \ if \ x \notin A \end{cases}.$$

(3.1)

On the other hand, a fuzzy set in a universe of discourse U is characterized by a MF $\mu_A(x)$, which may take any value from the interval of [0, 1] [32]. A fuzzy set may be represented using the following form:

$$A = \{(x, \mu_A(x)) | x \in U\}.$$

(3.2)

If U is continuous, the fuzzy set A can be represented by the following notation:

Fig. 3.1 A sample discrete
universe of discourse

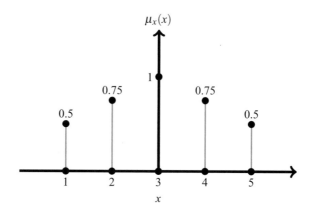

$$A_C = \int_U \frac{\mu_A(x)}{x}. \tag{3.3}$$

A typical $\mu_A(x)$ is illustrated in Fig. 3.1. When U is discrete, A is represented by the following notation:

$$A_D = \sum_U \frac{\mu_A(x)}{x}. \tag{3.4}$$

A discrete universe of discourse may be represented as in Fig. 3.1. In this case, the respective discrete universe of discourse may be represented as follows:

$$A = \left\{ (x, \mu_A(x)) | x \in X \right\} = \frac{0.5}{1} + \frac{0.75}{2} + \frac{1}{3} + \frac{0.75}{4} + \frac{0.5}{5}. \tag{3.5}$$

In this book, fuzzy sets with continuous universe of discourse are considered. There exist different types of type-1 MFs in literature.

3.1.2.1 Sigmoidal MFs

The shapes of Sigmoidal MFs depend on two parameters, a and m, defined in the following formulation:

$$f_{sig}(x, a, m) = \frac{1}{1 + e^{-a(x-m)}}. \tag{3.6}$$

3.1.2.2 Gaussian MFs

Gaussian MFs are one of the most frequently used MFs in literature and are defined as follows:

$$f_{Gauss}(x, \sigma, m) = e^{-\frac{(x-m)^2}{2\sigma^2}}, \tag{3.7}$$

where the parameter m represents the center of the MF and the parameter σ stands for the standard deviation of the MF.

3.1.2.3 Generalized Bell-Shaped MFs

Generalized bell-shaped MFs benefit from three parameters, a, b, and c. This MF is defined as follows:

$$f_{Gbell}(x, a, b, c) = \frac{1}{1 + \left|\frac{x-c}{a}\right|^{2b}}. \tag{3.8}$$

3.1.2.4 Trapezoidal Shaped MFs

Trapezoidal shaped MFs are defined as follows:

$$f_{Trap}(x, a, b, c, d) = max\left(min\left(\frac{x-a}{b-c}, 1, \frac{d-x}{c-b}\right), 0\right). \tag{3.9}$$

3.2 Type-1 Fuzzy Logic Systems

The inputs and the outputs of a T1FLS have crisp values. The duty of a typical T1FLS is to map the crisp inputs to the outputs using its rule base. The main parts of a T1FLS are the fuzzifier, the rule base, the inference engine, the defuzzifier, and the data base that the FLS is built upon [7]. These five main parts are illustrated in Fig. 3.2.

3.2.1 The Fuzzifier

This part maps the real-valued crisp inputs into type-1 fuzzy sets, which is further processed by the inference engine.

3.2.1.1 The Rule Base

The heart of a FLS is its rule base that consists of a set of fuzzy IF–THEN rules.

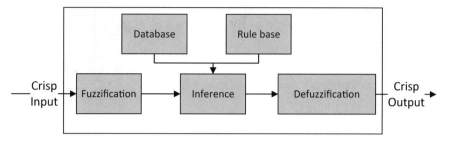

Fig. 3.2 Five main parts of a FLS

3.2.1.2 Inference Engine

An inference engine combines the fuzzy IF–THEN rules defined in rule base to map the fuzzy input set to the fuzzy output set. The product inference engine and minimum inference engine are among the most frequently used inference engines.

3.2.1.3 Defuzzifier

The defuzzification block defuzzifies the fuzzy output that is generated by the inference engine to produce the crisp output.

3.2.1.4 Data Base

Different data bases may result in the construction of different FLSs. Hence, it is one of the main parts of an FLS.

3.3 Type-2 Fuzzy Sets and Systems

A group of experts may propose different MFs for a single word. In other words, the MFs proposed by different experts may differ regarding some words like *tall, short, large, small*, and so on. Hence, the FLS constructed using human knowledge may have a histogram of MFs in the antecedent part and an interval of different values in its consequent part. A type-2 fuzzy set is a generalization of its type-1 counterpart whose membership grades are themselves fuzzy and benefit from a secondary membership grade that represent their validity.

A type-2 fuzzy set with a continuous universe of discourse is characterized by $\tilde{A}$ and is represented as follows:

$$\tilde{A} = \left\{ \left(x, u, \mu_{\tilde{A}}(x, u) \right) | \forall x \in X, \forall u \in J_x \subset [0, \ 1] \right\}, \tag{3.10}$$

where $0 \le \mu_{\tilde{A}}(x, u) \le 1$. The type-2 fuzzy set $\tilde{A}$ can also be defined in terms of the union of all admissible x's and u's as follows:

$$\tilde{A} = \int_{x \in X} \int_{u \in U} \mu_{\tilde{A}}(x, u)/(x, u) \ J_x \subseteq [0, \ 1]. \tag{3.11}$$

The secondary MF for an input is itself a fuzzy MF, which can be obtained for a single point x' as follows:

$$\mu_{\tilde{A}}(x = x', u) = \mu_{\tilde{A}}(x') = \int_{u \in J_{x'}} f_{x'}(u)/u \ J_{x'} \subseteq [0, \ 1], \tag{3.12}$$

where $0 \le f_{x'}(u) \le 1$ is the degree of MF corresponding to a single point x', which is itself another type-1 MF and is called the secondary MF. The FLS constructed based on such a MF is called a general type-2 fuzzy logic system.

It may be possible that all secondary MF values are taken as equal to either *zero* or *one*; the FLS in this case is called IT2FLS.

A Gaussian IT2MF with an uncertain center and a Gaussian IT2MF with an uncertain standard deviation are among the most frequently used IT2MFs. Particularly since Gaussian IT2MFs with uncertain standard deviations are differentiable in every points in their support set, they are more appropriate choice specially if the tuning of the antecedent part parameters using computational-based algorithms are intended. Existing IT2MFs are discussed in the next subsection.

The different parts of an IT2FLS are the fuzzifiers, the rule base, the inference engine, the type reducer, defuzzifier. The rules used in the structure of IT2FLSs are different from their type-1 counterpart in that they use interval MFs. A typical type-2 fuzzy rule is as follows:

$$If \ x_1 \ is \ \tilde{A}_1^l \ and \ \dots \ and \ x_p \ is \ \tilde{A}_p^l \ then \ y \ is \ \tilde{G}^l, \ \ l = 1, \dots, M, \tag{3.13}$$

where p is the number of inputs and M is the number of rules of the FLS. We know that for an IT2FLS that uses either minimum or product t-norm, the result of the input and antecedent operations is an interval type-1 set-the firing interval-which is determined by its left-most and right-most points [20]. The next step, after fuzzy inference, is type reduction with its result being an interval set as well.

This section introduces the structure of a MISO IT2FLS. While the input vector for the system is considered to be selected as $u = (u_1, \dots, u_{n_u})$, its output vector is taken as y.

1. Layer 0 (Input Layer): The inputs considered for the system are crisp values. The inputs considered for the system are fed to the fuzzy neural networks through this layer.
2. Layer 1 (MF Layer): Each node of this layer is an IT2MF. The jth MF for the ith input is represented by $\tilde{A}_{ij}$.
3. Layer 2 Nodes at second layer represent fuzzy rules. The outputs of this layer represent the *firing strength* of its corresponding rule. A *meet* operator connects IT2MFs together. The output of this layer is itself an interval value, which is calculated as follows:
4. Layer 3 (output processing layer): The calculation of the output of an IT2FLS with center-of-sets type-reduction is a nonlinear constrained optimization problem. This nonlinear constrained optimization problem is as follows [8].

3.3.1 Existing IT2MFs

As was mentioned earlier, the first layer of an IT2FLS is an MF Layer. Each node in this layer returns an interval that represents the left- and right-most points of an IT2MF.

Different kinds of IT2MFs exist in the literature, namely triangular, Gaussian, trapezoidal, sigmoidal, pi-shaped, etc. Gaussian IT2MFs, triangular IT2MFs, and elliptic IT2MF are investigated in this part.

The general mathematical expression for the Gaussian MF is expressed as follows:

$$\tilde{\mu}(x) = \exp\left(-\frac{1}{2}\frac{(x-c)^2}{\sigma^2}\right), \tag{3.14}$$

where c is the center of the MF, σ is the spread of the MF, and x is the input vector. It is possible to consider an interval for the center of the MFs. In this case, the parameter c can take any value from the interval of $[c_1, \ c_2]$. In this case, the right-most point of the MF is obtained as follows:

$$\overline{\mu}(x) = \begin{cases} \exp\left(-\frac{1}{2}\frac{(x-c_1)^2}{\sigma^2}\right) & if \ \ x < c_1 \\ 1 & if \ \ c_1 \le x \le c_2 \\ \exp\left(-\frac{1}{2}\frac{(x-c_2)^2}{\sigma^2}\right) & if \ \ c_2 < x \end{cases} \tag{3.15}$$

and its lower MF is obtained as follows:

$$\overline{\mu}(x) = \begin{cases} \exp\left(-\frac{1}{2}\frac{(x-c_2)^2}{\sigma^2}\right) & if \ \ x < (c_1 + c_2)/2 \\ \exp\left(-\frac{1}{2}\frac{(x-c_1)^2}{\sigma^2}\right) & if \ \ (c_1 + c_2)/2 \le x \end{cases} \tag{3.16}$$

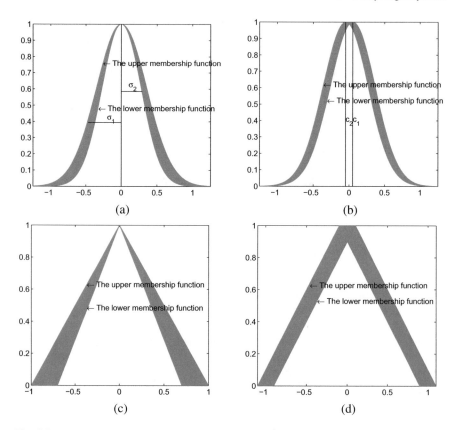

Fig. 3.3 Type-2 fuzzy set with **a** uncertain standard deviation and **b** uncertain mean. Triangular type-2 fuzzy set with **c** uncertain width and **d** uncertain center (**b**)

In Figs. 3.3a, b, Gaussian type-2 fuzzy sets with uncertain standard deviation and uncertain mean are illustrated.

Triangular IT2MFs with uncertain width and/or uncertain center are depicted in Fig. 3.3c, d. The mathematical expression for the MF is expressed as

$$\tilde{\mu}(x) = 1 - \frac{|x - c|}{d}, \tag{3.17}$$

where c and d are the center and the width of the MF, x is the input vector.

An elliptic IT2MF has certain values on both ends of the support and the kernel, and some uncertain values on the other parts of the support. This kind of MF is appropriate when complete belonging to a set and complete exclusion from a fuzzy set are fully certain but the degrees of membership in other parts of support are uncertain. The mathematical formula for such a MF is as follows:

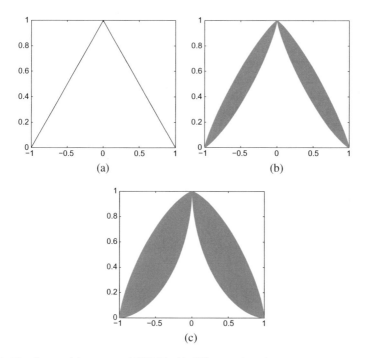

Fig. 3.4 The shapes of the proposed IT2MF with different values for a_1 and a_2

$$\tilde{\mu}(x) = \left(1 - \left|\frac{x-c}{d}\right|^a\right)^{\frac{1}{a}}, \quad a_2 < a < a_1, \quad c - d \le x \le c + d, \qquad (3.18)$$

where c is the center of the MF, and d is the spread of the MF, and x is the input vector. The parameters a_1 and a_2 determine the width of the uncertainty of the proposed MF. These parameters should be selected in the following form:

$$a_1 > 1 \qquad (3.19)$$
$$0 < a_2 < 1.$$

Figure 3.4a, b, c show the shapes of the elliptic IT2MF for $a_1 = a_2 = 1$, $a_1 = 1.2$, $a_2 = 0.8$ and $a_1 = 1.4$, $a_2 = 0.6$, respectively. As can be seen from Fig. 3.4a, this MF converges to a triangular one, when all of its uncertainties are removed, i.e., $a_1 = a_2 = 1$. However, if the values of a_1 and a_2 are selected as being different from *one*, the shape of this MF would be quite different from that of the triangular MF.

3.3.2 Output Processing Unit

In order to obtain the center-of-set type-reduction (COS TR) of IT2FLSs, y_l and y_r, the minimum value and the maximum value of the output of the interval output of the FLS, must be calculated then, the output of the FLS, y, is calculated as $y = \frac{y_l + y_r}{2}$. In order to obtain y_l and y_r, the following nonlinear constrained optimization processes must be solved [8].

$$y_l = \min_{\substack{\forall \theta_i \in [\underline{\theta}_i, \overline{\theta}_i] \\ \forall w_i \in [\underline{w}_i, \overline{w}_i]}} \frac{\sum_{i=1}^{N} \theta_i w_i}{\sum_{i=1}^{N} w_i}, \qquad (3.20)$$

$$y_r = \max_{\substack{\forall \theta_i \in [\underline{\theta}_i, \overline{\theta}_i] \\ \forall w_i \in [\underline{w}_i, \overline{w}_i]}} \frac{\sum_{i=1}^{N} \theta_i w_i}{\sum_{i=1}^{N} w_i} \qquad (3.21)$$

and

$$\theta_i \in \Theta_i \equiv [\underline{\theta}_i, \overline{\theta}_i], \quad i = 1, 2, \dots, N$$

$$w_i \in W_i \equiv [\underline{w}_i, \overline{w}_i], \quad i = 1, 2, \dots, N$$

where

$$\underline{\theta}_i \le \overline{\theta}_i \quad i = 1, 2, \dots, N$$

$$\underline{w}_i \le \overline{w}_i \quad i = 1, 2, \dots, N$$

and $\underline{\theta}_i$ and $\overline{\theta}_i$ are the left- and right-most points of the interval sets considered for the consequent part of ith rule of the FLS; $\underline{w}_i$ and $\overline{w}_i$ are left and right-most point of the firing of ith rule of the IT2FLS; and N is the number of the rules of the IT2FLS.

$$y_l = \min_{\substack{\forall x_i \in [\underline{x}_i, \overline{x}_i] \\ \forall w_i \in [\underline{w}_i, \overline{w}_i]}} \frac{\sum_{i=1}^{N} x_i w_i}{\sum_{i=1}^{N} w_i}, \qquad (3.22)$$

$$y_r = \max_{\substack{\forall x_i \in [\underline{x}_i, \overline{x}_i] \\ \forall w_i \in [\underline{w}_i, \overline{w}_i]}} \frac{\sum_{i=1}^{N} x_i w_i}{\sum_{i=1}^{N} w_i} \qquad (3.23)$$

are left- and right-most points of Y, respectively. The result of solving (3.20) and (3.21) can be expressed as

$$y_l = \frac{\sum_{i=1}^{L} x_i \overline{w}_i + \sum_{i=L+1}^{N} x_i \underline{w}_i}{\sum_{i=1}^{L} \overline{w}_i + \sum_{i=L+1}^{N} \underline{w}_i} \qquad (3.24)$$

and

$$y_r = \frac{\sum_{i=1}^{R} \overline{x}_i \underline{w}_i + \sum_{i=R+1}^{N} \overline{x}_i \overline{w}_i}{\sum_{i=1}^{R} \underline{w}_i + \sum_{i=R+1}^{N} \overline{w}_i}, \tag{3.25}$$

where L and R are switch points satisfying $\underline{x}_L \leq y_l < \underline{x}_{L+1}$ and $\overline{x}_R \leq y_r < \overline{x}_{R+1}$. As can be seen from the solution, the optimal solutions y_l and y_r either include $\underline{w}_i$ or $\overline{w}_i$. In other words, no intermediate value of w_i's exists in y_l and y_r. In order to find the solutions of (3.20) and (3.21), the switch points L and R must be determind. There exists no closed-form solutions to this problem.

3.3.3 Popular Existing Output Processing Units

As mentioned earlier, a number of alternative solutions to KM algorithms and approximate solutions to the problem formulated in the previous section have been proposed in literature. In this paper, we consider just the COS TR methods in the first category.

3.3.3.1 The Original Karnik–Mendel Method

3.3.3.2 KM Algorithm for Computing y_l [8]

I. Sort θ_i $(i = 1, \ldots, N)$ in increasing order so that $\theta_1 \leq \theta_2 \leq \cdots \leq \theta_N$. Match the weights w_i with their respective θ_i and renumber them to associate them with the reordered θ_i.

II. Initialize w_i by setting

$$w_i = \frac{\underline{w}_i + \overline{w}_i}{2} \quad i = 1, 2, \ldots, N \tag{3.26}$$

and then compute

$$y = \frac{\sum_{i=1}^{N} \theta_i w_i}{\sum_{i=1}^{N} w_i}. \tag{3.27}$$

III. Find switch point k $(1 \leq k \leq N - 1)$ such that

$$\theta_k \leq y \leq \theta_{k+1}. \tag{3.28}$$

IV. Set

$$w_i = \begin{cases} \overline{w}_i, & i \leq k \\ \underline{w}_i, & i > k \end{cases} \tag{3.29}$$

and compute

$$y' = \frac{\sum_{i=1}^{N} \theta_i w_i}{\sum_{i=1}^{N} w_i}. \tag{3.30}$$

V. Check if $y' = y$. If yes, stop, set $y_l = y$, and call k, L. If not, go to step VI.
VI. Set $y = y'$ and go to step III.

3.3.3.3   KM Algorithm for Computing y_r [8]

I. Sort $\overline{\theta}_i$ $(i = 1, \ldots, N)$ in increasing order so that $\overline{x}_1 \leq \overline{\theta}_2 \leq \cdots \leq \overline{\theta}_N$. Match the weights w_i with their respective $\overline{\theta}_i$ and renumber them to associate them with the reordered $\overline{\theta}_i$

II. Initialize w_i by setting

$$w_i = \frac{\underline{w}_i + \overline{w}_i}{2} \quad i = 1, 2, \ldots, N \tag{3.31}$$

and then compute

$$y = \frac{\sum_{i=1}^{N} \overline{\theta}_i w_i}{\sum_{i=1}^{N} w_i}. \tag{3.32}$$

III. Find switch point k $(1 \leq k \leq N - 1)$ such that

$$\overline{\theta}_k \leq y \leq \overline{\theta}_{k+1}. \tag{3.33}$$

IV. Set

$$w_i = \begin{cases} \underline{w}_i, & i \leq k \\ \overline{w}_i, & i > k \end{cases} \tag{3.34}$$

and compute

$$y' = \frac{\sum_{i=1}^{N} \overline{\theta}_i w_i}{\sum_{i=1}^{N} w_i}. \tag{3.35}$$

V. Check if $y' = y$. If yes, stop, set $y_l = y$, and call k, L. If not, go to step VI.
VI. Set $y = y'$ and go to step III.

3.3.3.4   Enhanced Karnik–Mendel Method

3.3.3.5   EKM Algorithm for Computing y_l [33]

I. Sort $\underline{\theta}_i$ $(i = 1, 2, \ldots, N)$ in increasing order and call the sorted $\underline{\theta}_i$ by the same name, but $\underline{\theta}_1 \leq \underline{\theta}_2 \leq \cdots \leq \underline{\theta}_N$. Match the weights w_i with their respective $\underline{\theta}_i$ and renumber them so that their index corresponds to the renumbered $\underline{\theta}_i$.

II. Set $k = [N/2.4]$ (the nearest integer to $N/2.4$), and compute

$$a = \sum_{i=1}^{k} \underline{\theta}_i \overline{w}_i + \sum_{i=k+1}^{N} \underline{\theta}_i \underline{w}_i$$

$$b = \sum_{i=1}^{k} \overline{w}_i + \sum_{i=k+1}^{N} \underline{w}_i$$

and

$$y = \frac{a}{b}.$$

III. Find $k' \in [1, \ N - 1]$ such that

$$\underline{\theta}_{k'} \leq y \leq \underline{\theta}_{k'+1}.$$

IV. Check if $k' = k$. If yes, stop, set $y_l = y$, and call k, L. If no, continue.
V. Compute $s = sign(k' - k)$, and

$$a' = a + s \sum_{i=min(k,k')+1}^{max(k,k')} \underline{\theta}_i (\overline{w}_i - \underline{w}_i),$$

$$b' = b + s \sum_{i=min(k,k')+1}^{max(k,k')} (\overline{w}_i - \underline{w}_i),$$

$$y' = \frac{a'}{b'}.$$

VI. Set $y = y'$, $a = a'$, $b = b'$ and $k = k'$. Go to step III.

3.3.3.6   EKM Algorithm for Computing y_r

The EKM algorithm for computing y_r is very similar to y_l and therefore, the reader is referred to [33] for the complete version of the algorithm.

1. Sort x_i $(i = 1, 2, \ldots, N)$ in increasing order and call the sorted x_i by the same name, but $\underline{\theta}_1 \leq \underline{\theta}_2 \leq \cdots \leq \underline{\theta}_N$. Match the weights w_i with their respective $\underline{\theta}_i$ and renumber them so that their index corresponds to the renumbered $\underline{\theta}_i$.
2. Set $k = [N/1.7]$ (the nearest integer to $N/1.7$), and compute

$$a = \sum_{i=1}^{k} \overline{\theta}_i \underline{w}_i + \sum_{i=k+1}^{N} \underline{\theta}_i \overline{w}_i, \tag{3.36}$$

$$b = \sum_{i=1}^{k} \underline{w}_i + \sum_{i=k+1}^{N} \overline{w}_i \tag{3.37}$$

and

$$y = \frac{a}{b}. \tag{3.38}$$

3. Find $k' \in [1, \; N-1]$ such that

$$\underline{\theta}_{k'} \leq y \leq \underline{\theta}_{k'+1}. \tag{3.39}$$

4. Check if $k' = k$. If yes, stop, set $y_r = y$, and call k, R. If no, continue.
5. Compute $s = sign(k' - k)$, and

$$a' = a - s \sum_{i=min(k,k')+1}^{max(k,k')} \underline{\theta}_i (\overline{w}_i - \underline{w}_i), \tag{3.40}$$

$$b' = b - s \sum_{i=min(k,k')+1}^{max(k,k')} (\overline{w}_i - \underline{w}_i), \tag{3.41}$$

$$y' = \frac{a'}{b'}. \tag{3.42}$$

6. Set $y = y'$, $a = a'$, $b = b'$ and $k = k'$. Go to step (3).

3.3.3.7   Iterative Algorithm with Stop Conditions

In [21], it is proved that the KM algorithms have the following properties:

– It converges monotonically and super-exponentially fast.
– If $y_l(k)$ is defined as follows, it monotonically decreases for $k <= L$ and then, monotonically increases with increase in k.

$$y_l(k) = \frac{\sum_{i=1}^{k} \underline{\theta}_i \overline{w}_i + \sum_{i=k+1}^{N} \underline{\theta}_i \underline{w}_i}{\sum_{i=1}^{k} \overline{w}_i + \sum_{i=k+1}^{N} \underline{w}_i}.$$

– If $y_r(k)$ is defined as follows, it monotonically increases for $k <= R$ and then it monotonically increases with increase in k.

$$y_r(k) = \frac{\sum_{i=1}^{k} \overline{\theta}_i \underline{w}_i + \sum_{i=k+1}^{N} \overline{\theta}_i \overline{w}_i}{\sum_{i=1}^{k} \underline{w}_i + \sum_{i=k+1}^{N} \overline{w}_i}.$$

Using these facts, Melgarejo and his co-authors have proposed an IASC to compute the generalized centroid of IT2FLSs that can also be used for the COS TR of IT2FLSs [5]. They increase the switching point (k) from 1 to $N-1$ and stop on the point at which $y_l(k)$ begins to change its direction (i.e., it increases instead of decreasing). They use the same concept for computing (R). They increase the switching point (k) from 1 to $N-1$ and stop on the point at which $y_r(k)$ begins to change its direction (i.e., it decreases instead of increasing). These algorithms are summarized as follows.

3.3.3.8 IASC for Computing y_l

I. The first step is the same as KM and EKM for computing y_l.
II. Initialize a, b, and y_l as follows:

$$a = \sum_{i=1}^{N} \underline{\theta}_i \overline{w}_i$$

$$b = \sum_{i=1}^{N} \overline{w}_i$$

$$y_l = \underline{\theta}_N$$

$$k = 0.$$

III. Compute

$$k = k + 1$$
$$a = a + \underline{\theta}_k (\overline{w}_k - \underline{w}_k)$$
$$b = b + \overline{w}_k - \underline{w}_k$$
$$c = a/b.$$

IV. If $c > y_l$, set $L = k - 1$ and $y_l = c$ and stop; otherwise, set $y_l = c$ and go to Step III.

3.3.3.9 IASC for Computing y_r

IASC algorithm for computing y_r is very similar to y_l. The reader is referred to [5, 19] for the complete version of the algorithm.

1. The first step is the same as KM and EKM for computing y_r.

2. Initialize a, b, and y_r as follows:

$$a = \sum_{i=1}^{N} \overline{\theta}_i \overline{w}_i \qquad (3.43)$$

$$b = \sum_{i=1}^{N} \overline{w}_i \qquad (3.44)$$

$$y_r = \overline{\theta}_1 \qquad (3.45)$$

$$k = 0. \qquad (3.46)$$

3. Compute

$$k = k + 1 \qquad (3.47)$$

$$a = a - \overline{\theta}_k(\overline{w}_k - \underline{w}_k) \qquad (3.48)$$

$$b = b - \overline{w}_k + \underline{w}_k \qquad (3.49)$$

$$c = a/b. \qquad (3.50)$$

4. If $c < y_r$, set $R = k - 1$ and $y_r = c$ and stop; otherwise, set $y_r = c$ and go to step 3.

3.3.3.10 Enhanced Iterative Algorithm with Stop Conditions

An enhanced version for IASC, in which two important modifications are proposed can be found in [34]. The first modification is the starting point for calculating y_r. In IASC, the initial switching point for y_r, (R) is taken as being equal to 1 and then, its true value is found by increasing the switching point by one and checking the stop conditions. However, it is known that the true value for R is close to $N/1.7$ [33], which is closer to N than to *one*. It then follows that the number of the required iterations may greatly decrease if the initial value for R is taken equal to N and its true value is found by decreasing the value of this switching point.

The second modification is a new stop condition for IASC algorithms. The new stop conditions are based on the following facts.

$$y_l(k) \begin{cases} \geq \underline{\theta}_k & k \leq L \\ < \underline{\theta}_k & k > L \end{cases},$$

$$y_r(k) \begin{cases} > \overline{\theta}_k & k \leq R \\ \leq \overline{\theta}_k & k > R \end{cases}$$

in which $y_l(k)$ and $y_r(k)$ are defined as follows:

$$y_l(k) = \frac{\sum_{i=1}^{k} \theta_i \overline{w}_i + \sum_{i=k+1}^{N} \theta_i \underline{w}_i}{\sum_{i=1}^{k} \overline{w}_i + \sum_{i=k+1}^{N} \underline{w}_i},$$

$$y_r(k) = \frac{\sum_{i=1}^{k} \overline{\theta}_i \underline{w}_i + \sum_{i=k+1}^{N} \overline{\theta}_i \overline{w}_i}{\sum_{i=1}^{k} \underline{w}_i + \sum_{i=k+1}^{N} \overline{w}_i}.$$

EIASC algorithms can be summarized as follows:

3.3.3.11 EIASC for Computing y_l

I. The first step is the same as EKM and KM for computing y_l.
II. Initialize a, b, y_l, and l as follows:

$$a = \sum_{i=1}^{N} \theta_i \underline{w}_i$$

$$b = \sum_{i=1}^{N} \underline{w}_i$$

$$y_l = \underline{\theta}_N$$

$$l = 0.$$

III. Compute

$$l = l + 1$$
$$a = a + \underline{\theta}_l(\overline{w}_l - \underline{w}_l)$$
$$b = b + \overline{w}_l - \underline{w}_l$$
$$c = a/b.$$

IV. If $c > y_l$, then set $L = l - 1$ and stop; otherwise, set $y_l = c$ and go to step III.

3.3.3.12 EIASC for Computing y_r

The EIASC algorithm for computing y_r is very similar to y_l. The reader is referred to [34] for the complete version of the algorithm.

76 3 Fuzzy Logic Systems

3.3.4 Center-of-Set Type-Reducer Without Sorting Requirement Algorithm

As can be seen from (3.20)–(3.21), in order to find the *min* and the *max* values of Y, each w_i must take the value equal to either $\underline{w}_i$ or $\overline{w}_i$. Therefore, (3.20) and (3.21) can be reformulated as follows:

$$y_l = \frac{\sum_{i=1}^{N} \lambda_{li}\overline{w}_i\underline{\theta}_i + \sum_{i=1}^{N}(1-\lambda_{li})\underline{w}_i\underline{\theta}_i}{\sum_{i=1}^{N} \lambda_{li}\overline{w}_i + \sum_{i=1}^{N}(1-\lambda_{li})\underline{w}_i}, \quad \lambda_{li} \in [0, \ 1] \quad (3.51)$$

$$y_r = \frac{\sum_{i=1}^{N} \lambda_{ri}\overline{w}_i\overline{\theta}_i + \sum_{i=1}^{N}(1-\lambda_{ri})\underline{w}_i\overline{\theta}_i}{\sum_{i=1}^{N} \lambda_{ri}\overline{w}_i + \sum_{i=1}^{N}(1-\lambda_i)\underline{w}_{ri}}, \quad \lambda_{ri} \in [0, \ 1] \quad (3.52)$$

where λ_{li} and λ_{ri} can take the any values from the interval of $[0, \ 1]$. However, if the final values of λ_{li} and λ_{ri} are taken as either equal to *zero* or *one*, the resulting formula for y_l and y_r are two alternative algorithms for KM. In the new formulation, L and R do not exist any more and we must just determine which of the λ_{ri}'s and the λ_{li}'s must be *one* and which ones of them must be *zero*. As can be seen from (3.62), if $\lambda_{li} = 1$, it means that $\overline{w}_i$ is selected for w_i. In the case when $\lambda_{li} = 0$ $\underline{w}_i$ is selected for w_i. A similar discussion can be made for λ_{ri}'s.

3.3.4.1 The Summary of the Proposed Algorithm for Computing y_l

The proposed algorithm for computing y_l is summarized as follows:

I. If $\underline{w}_i = 0, \ i = 1, \ldots, N$, then y_l is equal to the minimum value of $\underline{\theta}_i$'s whose $\overline{w}_i \neq 0, i = 1, \ldots, N$ and stop.
II. Initialize $\lambda_{li} = 0.5, \ i = 1, \ldots, N$
III. Calculate $\delta_1, \delta_{2l}, \delta_{3l}, \delta_{4l}$ as follows:

$$\delta_1 = \sum_{i=1}^{N} \overline{w}_i, \quad (3.53)$$

$$\delta_{2l} = \sum_{i=1}^{N} \overline{w}_i\underline{\theta}_i, \quad \delta_{2r} = \sum_{i=1}^{N} \overline{w}_i\overline{\theta}_i, \quad (3.54)$$

$$\delta_{3l} = \sum_{i=1}^{N} \Delta w_i (1 - \lambda_{li}) \underline{\theta}_i, \quad \delta_{3r} = \sum_{i=1}^{N} \Delta w_i (1 - \lambda_{ri}) \overline{\theta}_i, \qquad (3.55)$$

$$\delta_{4l} = \sum_{i=1}^{N} \Delta w_i (1 - \lambda_{li}), \quad \delta_{4r} = \sum_{i=1}^{N} \Delta w_i (1 - \lambda_{ri}). \qquad (3.56)$$

IV. $flag = 0$

V. For $j = 1$ to N:

{Calculate A_{lj} as follows:

$$A_{lj} = \underline{\theta}_j - \frac{\delta_{2l}}{\delta_1} + \frac{\delta_{3l}}{\delta_1} - \frac{\delta_{4l}}{\delta_1} \underline{\theta}_j \qquad (3.57)$$

if $A_{lj} < 0$, then $\lambda'_{lj} = 1$ else $\lambda'_{lj} = 0$

if $\lambda_{lj} \neq \lambda'_{lj}$ then update $flag, \delta_{3l}, \delta_{4l}$, and λ_{lj} as follows:

$$\{flag = 1$$
$$\delta_{3l} = \delta_{3l} + \Delta w_j \underline{\theta}_j (\lambda_{lj} - \lambda'_{lj})$$
$$\delta_{4l} = \delta_{4l} + \Delta w_j (\lambda_{lj} - \lambda'_{lj})$$
$$\lambda_{lj} = \lambda'_{lj} \}.$$

}

VI. If $flag = 0$, then calculate y_l as

$$y_l = \frac{\delta_{2l} - \delta_{3l}}{\delta_{1l} - \delta_{4l}} \qquad (3.58)$$

and stop, else go to step III.

3.3.4.2 The Summary of the Proposed Algorithm for Computing y_r

The proposed algorithm for computing y_r is summarized as follows:

I. If $\underline{w}_i = 0$, $i = 1, \ldots, N$, then y_r is equal to the maximum value of $\overline{\theta}_i$'s, whose $\overline{w}_i \neq 0, i = 1, \ldots, N$ and stop.

II. Initialize $\lambda_{ri} = 0.5$, $i = 1, \ldots, N$

III. Calculate $\delta_1, \delta_{2r}, \delta_{3r}, \delta_{4r}$ as in (3.53)–(3.56)

IV. $flag = 0$

V. For $j = 1$ to N:

{Calculate A_{rj} as follows:

$$A_{rj} = \overline{\theta}_j - \frac{\delta_{2r}}{\delta_1} + \frac{\delta_{3r}}{\delta_1} - \frac{\delta_{4r}}{\delta_1}\underline{\theta}_j \qquad (3.59)$$

if $A_{rj} > 0$, then $\lambda'_{rj} = 1$ else $\lambda'_{rj} = 0$
if $\lambda_{rj} \neq \lambda'_{rj}$, then update $flag, \delta_{3r}, \delta_{4r}$ and λ_{rj} as follows:

$$\begin{aligned}
\{flag &= 1\\
\delta_{3r} &= \delta_{3r} + \Delta w_j \overline{\theta}_j (\lambda_{rj} - \lambda'_{rj})\\
\delta_{4r} &= \delta_{4r} + \Delta w_j (\lambda_{rj} - \lambda'_{rj})\\
\lambda_{rj} &= \lambda'_{rj}\}.
\end{aligned}$$

}
VI. If $flag = 0$, then calculate y_r as:

$$y_r = \frac{\delta_{2r} - \delta_{3r}}{\delta_{1r} - \delta_{4r}} \qquad (3.60)$$

and stop, else go to Step III.

It is to be noted that these algorithms have been previously presented in [11, 12]. However, this paper includes newer examples and comparison with the six other existing variants of the KM method.

3.3.5   Family of Non-Iterative Output Processing Units

In order to analyze the stability of the identification system and the controller designed based on IT2FLS, the input/output relationship of the system must be explicitly described. It would reduce the complexity of the IT2FLS and make it possible to analyze its stability.

3.3.5.1   BMM

As this defuzzifier has a closed form, it is possible to investigate the stability analysis of the IT2FLS when acting as a controller [3] and/or identifier [9]. The output of an IT2FLS in a closed form can be approximated using the BMM method as follows:

$$Y_{BMM} = q\frac{\sum_1^N \overline{w}_i\theta_i}{\overline{w}_i} + (1 - q)\frac{\sum_1^N \underline{w}_i\theta_i}{\underline{w}_i}, \qquad (3.61)$$

where the parameter q can be selected from the interval [0, 1].

The BMM method can be further analyzed to show how well this algorithm is approximating the KM method. For this purpose, the original constrained optimization problem needs to be reformulated. As can be observed from (3.20) to (3.21), the minimum and the maximum values of Y happen exactly in the cases where each w_i is either equal to $\underline{w}_i$ or $\overline{w}_i$. Considering this fact, (3.24) and (3.25) can be reformulated as follows [15]:

$$
y_l = \frac{\sum_{i=1}^{N} \lambda_{li} \overline{w}_i \underline{\theta}_i + \sum_{i=1}^{N} (1 - \lambda_{li}) \underline{w}_i \underline{\theta}_i}{\sum_{i=1}^{N} \lambda_{li} \overline{w}_i + \sum_{i=1}^{N} (1 - \lambda_{li}) \underline{w}_i}, \quad \lambda_{li} \in [0, \ 1] \tag{3.62}
$$

$$
y_r = \frac{\sum_{i=1}^{N} \lambda_{ri} \overline{w}_i \overline{\theta}_i + \sum_{i=1}^{N} (1 - \lambda_{ri}) \underline{w}_i \overline{\theta}_i}{\sum_{i=1}^{N} \lambda_{ri} \overline{w}_i + \sum_{i=1}^{N} (1 - \lambda_{ri}) \underline{w}_i}, \quad \lambda_{ri} \in [0, \ 1], \tag{3.63}
$$

where λ_{li} and λ_{ri} can take any values from the interval of $[0, \ 1]$. However, if the final values of λ_{li} and λ_{ri} are taken as either equal to *zero* or *one*, the resulting formula for y_l and y_r are two alternative algorithms for the KM method. In the new formulation, L and R do not exist any more and we must just determine which of the λ_{ri}s and the λ_{li}s must be *one* and which ones of them must be *zero*. As can be seen from (3.51), if $\lambda_{li} = 1$, it means that $\overline{w}_i$ is selected for w_i. When $\lambda_{li} = 0$, $\underline{w}_i$ is selected for w_i. A similar discussion can be made for λ_{ri}s.

Here, in order to be able to take the derivative of y_l and y_r with respect to these parameters, λ_{li}s and λ_{ri}s are taken as any real number between *zero* and *one* rather than two binary values.

$$
y_l = \frac{\sum_{i=1}^{N} \lambda_{li} \overline{w}_i \underline{\theta}_i + \sum_{i=1}^{N} (1 - \lambda_{li})(\overline{w}_i - \Delta w_i)\underline{\theta}_i}{\sum_{i=1}^{N} \lambda_{li} \overline{w}_i + \sum_{i=1}^{N} (1 - \lambda_{li})(\overline{w}_i - \Delta w_i)}, \tag{3.64}
$$

where

$$
\Delta w_i = \overline{w}_i - \underline{w}_i.
$$

So that

$$
y_l = \frac{\sum_{i=1}^{N} \overline{w}_i \underline{\theta}_i - \sum_{i=1}^{N} (1 - \lambda_{li}) \Delta w_i \underline{\theta}_i}{\sum_{i=1}^{N} \overline{w}_i - \sum_{i=1}^{N} (1 - \lambda_{li}) \Delta w_i} \tag{3.65}
$$

and further

$$y_l = \frac{1}{\displaystyle\sum_{i=1}^{N} \overline{w}_i} \frac{\displaystyle\sum_{i=1}^{N} \overline{w}_i \underline{\theta}_i - \sum_{i=1}^{N} (1 - \lambda_{li}) \Delta w_i \underline{\theta}_i}{1 - \dfrac{\displaystyle\sum_{i=1}^{N} (1 - \lambda_{li}) \Delta w_i}{\displaystyle\sum_{i=1}^{N} \overline{w}_i}}. \tag{3.66}$$

Lemma 3.1 *The Maclaurin expansion of $1/(1+x)$ is as follows:*

$$\frac{1}{1+x} = 1 - x + x^2 - x^3 + \cdots, \quad |x| < 1. \tag{3.67}$$

Using the Maclaurin expansion as in Lemma 3.1, the following equation for y_l is obtained.

$$y_l = \frac{1}{\displaystyle\sum_{i=1}^{N} \overline{w}_i} \left(\sum_{i=1}^{N} \overline{w}_i \underline{\theta}_i - \sum_{i=1}^{N} (1 - \lambda_{li}) \Delta w_i \underline{\theta}_i \right) \times$$

$$\left(1 + \frac{\sum_{i=1}^{N} (1 - \lambda_{li}) \Delta w_i}{\sum_{i=1}^{N} \overline{w}_i} \right.$$

$$\left. + \left(\frac{\sum_{i=1}^{N} (1 - \lambda_{li}) \Delta w_i}{\sum_{i=1}^{N} \overline{w}_i} \right)^2 + H.O.T. \right). \tag{3.68}$$

The constrain required for this Maclaurin expansion is as follows:

$$\left| \frac{\displaystyle\sum_{i=1}^{N} (1 - \lambda_{li}) \Delta w_i}{\displaystyle\sum_{i=1}^{N} \overline{w}_i} \right| < 1. \tag{3.69}$$

Equation (3.69) is already satisfied as $\Delta w_i < \overline{w}_i$ and $(1 - \lambda_{li})$ is either *zero* or *one*.

The rest of the analysis is based on the first-order approximate Maclaurin series expansion. This approximation is closer to the exact solution if a smaller footprint of uncertainty does exist in the IT2MFs.

If the higher-order terms in (3.68) are neglected, the following first-order approximate of y_l is obtained.

$$y_l \approx \frac{1}{\displaystyle\sum_{i=1}^{N} \overline{w}_i} \left(\sum_{i=1}^{N} \overline{w}_i \underline{\theta}_i - \sum_{i=1}^{N} (1 - \lambda_{li}) \Delta w_i \underline{\theta}_i \right.$$

$$\left. + \frac{\displaystyle\sum_{i=1}^{N} \overline{w}_i \underline{\theta}_i \sum_{i=1}^{N} (1 - \lambda_{li}) \Delta w_i}{\displaystyle\sum_{i=1}^{N} \overline{w}_i} \right). \tag{3.70}$$

It is further possible to neglect the terms, including Δw_i, to obtain *zero*-order approximate for y_l as follows:

$$y_l \approx \frac{1}{\displaystyle\sum_{i=1}^{N} \overline{w}_i} \left(\sum_{i=1}^{N} \overline{w}_i \underline{\theta}_i \right). \tag{3.71}$$

It is possible to use the same Maclaurin series expansion as the one used to find the first-order polynomial approximation value for y_l, the following first-order approximation for y_r is obtained.

$$y_r \approx \frac{1}{\displaystyle\sum_{i=1}^{N} \overline{w}_i} \left(\sum_{i=1}^{N} \overline{w}_i \overline{\theta}_i - \sum_{i=1}^{N} (1 - \lambda_{ri}) \Delta w_i \overline{\theta}_i \right.$$

$$\left. + \frac{\displaystyle\sum_{i=1}^{N} \overline{w}_i \overline{\theta}_i \sum_{i=1}^{N} (1 - \lambda_{ri}) \Delta w_i}{\displaystyle\sum_{i=1}^{N} \overline{w}_i} \right). \tag{3.72}$$

Similarly, the *zero*-order approximation of y_r is obtained by neglecting the terms including Δw_i as follows:

$$y_r \approx \frac{1}{\displaystyle\sum_{i=1}^{N} \overline{w}_i} \left(\sum_{i=1}^{N} \overline{w}_i \overline{\theta}_i \right). \tag{3.73}$$

However, BMM is mainly designed for the case where $\overline{\theta}_i = \underline{\theta}_i = \theta_i$. In summary, we have

$$y_l = \frac{\sum\limits_{i=1}^{N} \underline{w}_i \theta_i}{\sum\limits_{i=1}^{N} \underline{w}_i}, \tag{3.74}$$

$$y_r = \frac{\sum\limits_{i=1}^{N} \overline{w}_i \theta_i}{\sum\limits_{i=1}^{N} \overline{w}_i}. \tag{3.75}$$

Considering (3.75) and (3.74), in both cases, we have the following equation as an approximate for the output of IT2FLS:

$$y = \frac{1}{2}\left(\frac{\sum\limits_{i=1}^{N} \overline{w}_i \theta_i}{\sum\limits_{i=1}^{N} \overline{w}_i} + \frac{\sum\limits_{i=1}^{N} \underline{w}_i \theta_i}{\sum\limits_{i=1}^{N} \underline{w}_i} \right). \tag{3.76}$$

From (3.61) and (3.76), it can be observed that the BMM algorithm is a zero-order approximate center of sets defuzzifier + TR when $m = n = 0.5$.

3.3.5.2 Nie–Tan Type-Reducer

Other than the BMM algorithm, the NT algorithm may also be used to reduce the complexity of KM and its variants to make them more suitable for real-time applications. The use of this method made it possible to analyze the noise reduction property of IT2FLSs. The output of a IT2FLS using NT is as follows:

$$y = \frac{\sum_{i=1}^{N} \theta_i \left(\underline{w}_i + \overline{w}_i \right)}{\sum_{i=1}^{N} \underline{w}_i + \overline{w}_i}. \tag{3.77}$$

3.3.5.3 Enhanced BMM Algorithm: The First-Order Approximation of KM Algorithms

As was mentioned earlier, BMM is a *zero*-order approximate polynomial type-reducer. In order to obtain a more exact approximate polynomial type-reducer, it is possible not to neglect the terms including first-order power of Δw_i. The first-order approximate polynomial function for y_l is as follows:

$$y_l \approx \frac{1}{\sum\limits_{i=1}^{N} \overline{w}_i} \left(\sum_{i=1}^{N} \overline{w}_i \underline{\theta}_i - \sum_{i=1}^{N} (1 - \lambda_{li}) \Delta w_i \underline{\theta}_i \right.$$

$$+ \left. \frac{\sum\limits_{i=1}^{N} \overline{w}_i \underline{\theta}_i \sum_{i=1}^{N} (1 - \lambda_{li}) \Delta w_i}{\sum\limits_{i=1}^{N} \overline{w}_i} \right). \tag{3.78}$$

The derivative of approximate function of y_l with respect to λ_{li} is as follows:

$$\frac{\partial y_l}{\lambda_{li}} \approx \frac{\Delta w_i}{\sum\limits_{i=1}^{N} \overline{w}_i} \left(\underline{\theta}_i - \frac{\sum\limits_{i=1}^{N} \overline{w}_i \underline{\theta}_i}{\sum\limits_{i=1}^{N} \overline{w}_i} \right). \tag{3.79}$$

Since the sign of $\frac{\partial y_l}{\lambda_{li}}$, in this case, depends on the sign of $\left(\underline{\theta}_i - \frac{\sum_{i=1}^{N} \overline{w}_i \underline{\theta}_i}{\sum_{i=1}^{N} \overline{w}_i} \right)$, and the fact that λ_{li} can be either *zero* or *one*, in order to obtain the first-order approximate of y_l, we have the following selections for λ_{li}:

$$\lambda_{li} = \begin{cases} 0 & \underline{m}_i > 0 \\ 1 & \underline{m}_i < 0 \end{cases} \tag{3.80}$$

or

$$\lambda_{li} = \frac{1 - sgn(\underline{m}_i)}{2} \tag{3.81}$$

where

$$\underline{m}_i = \underline{\theta}_i - \frac{\sum\limits_{i=1}^{N} \overline{w}_i \underline{\theta}_i}{\sum\limits_{i=1}^{N} \overline{w}_i}. \tag{3.82}$$

Applying (3.81)–(3.51), we have the following equation as the first-order approximate of y_l:

$$y_{l,proposed} \approx \frac{\sum\limits_{i=1}^{N} (\overline{w}_i + \underline{w}_i) \underline{\theta}_i - \sum\limits_{i=1}^{N} (sign(\underline{m}_i) \Delta w_i \underline{\theta}_i)}{\sum\limits_{i=1}^{N} (\overline{w}_i + \underline{w}_i) - \sum\limits_{i=1}^{N} (sign(\underline{m}_i) \Delta w_i)}. \tag{3.83}$$

Similarly, when the first-order terms, including Δw_i in (3.72) are taking into the account in the calculation of y_r, the following equation is obtained.

$$y_r \approx \frac{1}{\sum\limits_{i=1}^{N} \overline{w}_i} \left(\sum_{i=1}^{N} \overline{w}_i \overline{\theta}_i - \sum_{i=1}^{N} (1 - \lambda_{ri}) \Delta w_i \overline{\theta}_i \right.$$

$$\left. + \frac{\sum\limits_{i=1}^{N} \overline{w}_i \overline{\theta}_i \sum_{i=1}^{N} (1 - \lambda_{ri}) \Delta w_i}{\sum\limits_{i=1}^{N} \overline{w}_i} \right). \tag{3.84}$$

In this case, the derivative of the first-order polynomial approximate of y_r with respect to λ_{ri} is as follows:

$$\frac{\partial y_r}{\lambda_{ri}} \approx \frac{\Delta w_i}{\sum\limits_{i=1}^{N} \overline{w}_i} \left(\overline{\theta}_i - \frac{\sum\limits_{i=1}^{N} \overline{w}_i \overline{\theta}_i}{\sum\limits_{i=1}^{N} \overline{w}_i} \right). \tag{3.85}$$

Since the sign of $\frac{\partial y_r}{\lambda_{ri}}$ in this case depends on the sign of $\left(\overline{\theta}_i - \frac{\sum\limits_{i=1}^{N} \overline{w}_i \overline{\theta}_i}{\sum\limits_{i=1}^{N} \overline{w}_i} \right)$, and the fact that λ_{ri} can be either *zero* or *one*, in order to obtain the first-order approximate of y_r, we have the following selections for λ_{li}:

$$\lambda_{ri} = \begin{cases} 1 & \overline{m}_i > 0 \\ 0 & \overline{m}_i < 0 \end{cases} \tag{3.86}$$

or

$$\lambda_{ri} = \frac{sgn(\overline{m}_i) + 1}{2} \tag{3.87}$$

where

$$\overline{m}_i = \overline{\theta}_i - \frac{\sum\limits_{i=1}^{N} \overline{w}_i \overline{\theta}_i}{\sum\limits_{i=1}^{N} \overline{w}_i}. \tag{3.88}$$

By replacing (3.87) in (3.63), the following equation is obtained for the first-order approximate of y_r:

$$y_{r, proposed} \approx \frac{\sum\limits_{i=1}^{N} (\overline{w}_i + \underline{w}_i) \overline{\theta}_i + \sum\limits_{i=1}^{N} (sign(\overline{m}_i) \Delta w_i \overline{\theta}_i)}{\sum\limits_{i=1}^{N} (\overline{w}_i + \underline{w}_i) + \sum\limits_{i=1}^{N} (sign(\overline{m}_i) \Delta w_i)}. \tag{3.89}$$

It is to be noted that it is possible to add a correction term to the BMM method for y_l and y_r to obtain the complete first-order polynomial approximate for y_l and y_r.

$$y_{r,proposed} = y_{r,BMM} + \frac{A_r}{B_r} \tag{3.90}$$

in which

$$A_r = \sum_{i=1}^{N} \underline{w}_i \theta_i \sum_{i=1}^{N} \overline{w}_i + \sum_{i=1}^{N} sign(\overline{m}_i)\underline{w}_i \sum_{i=1}^{N} \overline{w}_i \theta_i$$

$$- \sum_{i=1}^{N} sign(\overline{m}_i)\underline{w}_i \theta_i \sum_{i=1}^{N} \overline{w}_i - \sum_{i=1}^{N} \overline{w}_i \theta_i \sum_{i=1}^{N} \underline{w}_i \tag{3.91}$$

and

$$B_r = \sum_{i=1}^{N} sign(\overline{m}_i)\underline{w}_i \sum_{i=1}^{N} \overline{w}_i - \sum_{i=1}^{N} sign(\overline{m}_i)\underline{w}_i \sum_{i=1}^{N} \overline{w}_i. \tag{3.92}$$

Similarly, we have

$$y_{l,proposed} = y_{l,BMM} + \frac{A_l}{B_l} \tag{3.93}$$

in which

$$A_l = \sum_{i=1}^{N} \overline{w}_i \theta_i \sum_{i=1}^{N} \underline{w}_i + \sum_{i=1}^{N} sign(\underline{m}_i)\overline{w}_i \theta_i \sum_{i=1}^{N} \underline{w}_i$$

$$+ \sum_{i=1}^{N} sign(\underline{m}_i)\overline{w}_i \sum_{i=1}^{N} \overline{w}_i \theta_i - \sum_{i=1}^{N} \overline{w}_i \theta_i \sum_{i=1}^{N} \underline{w}_i \tag{3.94}$$

and

$$B_l = \sum_{i=1}^{N} sign(\underline{m}_i)\overline{w}_i \sum_{i=1}^{N} \underline{w}_i + \sum_{i=1}^{N} sign(\underline{m}_i)\overline{w}_i \sum_{i=1}^{N} \overline{w}_i. \tag{3.95}$$

References

1. Åström, K.J., Wittenmark, B.: Adaptive Control. Courier Corporation, North Chelmsford (2013)
2. Åström, K.J., Hägglund, T., Astrom, K.J.: Advanced PID control (2006)

3. Biglarbegian, M., Melek, W.W., Mendel, J.M.: On the stability of interval type-2 tsk fuzzy logic control systems. IEEE Trans. Syst. Man Cybern. Part B (Cybern.) **40**(3), 798–818 (2010)
4. Chen, Y., Atherton, D.P., et al.: Linear Feedback Control: Analysis and Design with MATLAB, vol. 14. SIAM, Philadelphia (2007)
5. Duran, K., Bernal, H., Melgarejo, M.: Improved iterative algorithm for computing the generalized centroid of an interval type-2 fuzzy set. In: Fuzzy Information Processing Society, 2008. NAFIPS 2008. Annual Meeting of the North American, pp. 1–5. IEEE (2008)
6. Er, M.J., Liew, K.C.: Control of adept one SCARA robot using neural networks. IEEE Trans. Ind. Electron. **44**(6), 762–768 (1997)
7. Hoffmann, F.: Evolutionary algorithms for fuzzy control system design. Proc. IEEE **89**(9), 1318–1333 (2001)
8. Karnik, N.N., Mendel, J.M.: Centroid of a type-2 fuzzy set. Inf. Sci. **132**(1), 195–220 (2001)
9. Kayacan, E., Kayacan, E., Khanesar, M.A.: Identification of nonlinear dynamic systems using type-2 fuzzy neural networks-a novel learning algorithm and a comparative study. IEEE Trans. Ind. Electron. **62**(3), 1716–1724 (2015)
10. Khalil, H.K., Grizzle, J.: Nonlinear Systems, vol. 3. Prentice Hall, Upper Saddle River (2002)
11. Khanesar, M.A., Kaynak, O., Gao, H.: Improved Karnik-Mendel algorithm: eliminating the need for sorting. In: 2014 International Conference on Mechatronics and Control (ICMC), July, pp. 204–209 (2014)
12. Khanesar, M.A., Jalalian, A., Kaynak, O., Gao, H.: Improving the speed of center of set type-reduction in interval type-2 fuzzy systems by eliminating the need for sorting. IEEE Trans. Fuzzy Syst. (2016)
13. Khanesar, M.A., Kayacan, E., Teshnehlab, M., Kaynak, O.: Extended Kalman filter based learning algorithm for type-2 fuzzy logic systems and its experimental evaluation. IEEE Trans. Ind. Electron. **59**(11), 4443–4455 (2012)
14. Khanesar, M.A., Kaynak, O., Teshnehlab, M.: Direct model reference Takagi-Sugeno fuzzy control of SISO nonlinear systems. IEEE Trans. Fuzzy Syst. **19**(5), 914–924 (2011)
15. Khanesar, M.A., Khakshour, A.J., Kaynak, O., Gao, H.: Improving the speed of center of sets type reduction in interval type-2 fuzzy systems by eliminating the need for sorting. IEEE Trans. Fuzzy Syst. **25**(5), 1193–1206 (2017)
16. Liu, D., Yi, J., Zhao, D., Wang, W.: Adaptive sliding mode fuzzy control for a two-dimensional overhead crane. Mechatronics **15**(5), 505–522 (2005)
17. Mamdani, E.H., Assilian, S.: An experiment in linguistic synthesis with a fuzzy logic controller. Int. J. Man-Mach. Stud. **7**(1), 1–13 (1975)
18. Mayr, O.: The Origins Of Feedback Control. MIT Press, Cambridge (1975)
19. Melgarejo, M.: A fast recursive method to compute the generalized centroid of an interval type-2 fuzzy set. In: Fuzzy Information Processing Society, 2007. NAFIPS'07. Annual Meeting of the North American, pp. 190–194. IEEE (2007)
20. Mendel, J.M.: Uncertain rule-based fuzzy systems. In: Introduction and New Directions, p. 684. Springer, Berlin (2017)
21. Mendel, J.M., Liu, F.: Super-exponential convergence of the Karnik-Mendel algorithms for computing the centroid of an interval type-2 fuzzy set. IEEE Trans. Fuzzy Syst. **15**(2), 309–320 (2007)
22. Misir, D., Malki, H.A., Chen, G.: Design and analysis of a fuzzy proportional-integral-derivative controller. Fuzzy Sets Syst. **79**(3), 297–314 (1996)
23. Miyamura, A., Kimura, H.: Stability of feedback error learning scheme. Syst. Control Lett. **45**(4), 303–316 (2002)
24. Nakanishi, J., Schaal, S.: Feedback error learning and nonlinear adaptive control. Neural Netw. **17**(10), 1453–1465 (2004)
25. Ruan, X., Ding, M., Gong, D., Qiao, J.: On-line adaptive control for inverted pendulum balancing based on feedback-error-learning. Neurocomputing **70**(4–6), 770–776 (2007)
26. Ruan, X., Liu, L., Yu, N., Ding, M.: A model of feedback error learning based on Kalman estimator. In: 6th World Congress on Intelligent Control and Automation, 2006. WCICA 2006, vol. 1, pp. 4190–4194. IEEE (2006)

27. Sabahi, K., Teshnehlab, M., et al.: Recurrent fuzzy neural network by using feedback error learning approaches for LFC in interconnected power system. Energy Convers. Manag. **50**(4), 938–946 (2009)
28. Shaobai, Z., Xiaogang, R., XieFeng, C.: A new cerebellar learning scheme of feedback error based on Kalman estimator. In: Chinese Control and Decision Conference, 2008. CCDC 2008, pp. 4064–4069. IEEE (2008)
29. Shaocheng, T., Changying, L., Yongming, L.: Fuzzy adaptive observer backstepping control for MIMO nonlinear systems. Fuzzy Sets Syst. **160**(19), 2755–2775 (2009)
30. Sugeno, M., Nishida, M.: Fuzzy control of model car. Fuzzy Sets Syst. **16**(2), 103–113 (1985)
31. Tanaka, K., Wang, H.O.: Fuzzy control systems design and analysis: a linear matrix inequality approach. Wiley, Hoboken (2004)
32. Wang, L.X.: A course in fuzzy systems. Prentice-Hall Press, Upper Saddle River (1999)
33. Wu, D., Mendel, J.M.: Enhanced Karnik-Mendel algorithms. IEEE Trans. Fuzzy Syst. **17**(4), 923–934 (2009)
34. Wu, D., Nie, M.: Comparison and practical implementation of type-reduction algorithms for type-2 fuzzy sets and systems. In: 2011 IEEE International Conference on Fuzzy Systems (FUZZ), pp. 2131–2138. IEEE (2011)
35. Ying, H.: Sufficient conditions on general fuzzy systems as function approximators. Automatica **30**(3), 521–525 (1994)
36. Zadeh, L.: Fuzzy sets. Inf. Control **8**(3), 338–353 (1965)
37. Zadeh, L.A.: Is there a need for fuzzy logic? Inf. Sci. **178**(13), 2751–2779 (2008)

Chapter 4
Rule-Based Sliding-Mode Fuzzy Logic Control

4.1 Introduction

Fuzzy logic has proved itself as an advanced model-free approach with a tremendous impact on control community. Fuzzy logic has the ability to handle uncertainties, lack of modeling, and operational disturbances in a control system using expert knowledge. In light of the aforementioned abilities, fuzzy logic has the potential to be used as a complementary tool to overcome the shortcomings and challenges of sliding-mode controllers.

One of the main drawbacks of sliding-mode controllers is the chattering phenomena, which is caused by the $sign$ function used in their design. An additional problem with the use of the sign function in the control signal is that a noise of smaller amplitude acting on the sliding manifold may change its direction completely, and push system states in an inappropriate direction.

Less chattering in control signal decreases the switching frequency considerably and makes the system, in addition to other advantages, implementable using cost-effective actuators that can operate on lower frequencies. Since there is always some noise that is detected by sensors and corrupts the collected data from the process, it is highly desirable to lessen the sensitivity of the control signal to noise.

The existing fuzzy logic-based modifications to sliding-mode controllers can be mainly put into two categories: direct approaches and indirect approaches.

(1) Indirect Approaches: In order to overcome chattering and sensitivity to noise problems, fuzzy logic may be utilized. In this case, fuzzy logic theory may be used to tune sliding-mode controller parameters to overcome the aforementioned shortcomings. The typical parameters of the sliding-mode controller, which may be tuned using fuzzy logic, are the width of the saturation function and its time derivative or the distance between system states and the sliding manifold.

(2) Direct Approaches: In this case, the sliding-mode controller is completely replaced with a number of fuzzy IF–THEN rules. Therefore, the traditional procedure of the sliding-mode controller design, which may require more knowledge about

© Springer Nature Switzerland AG 2021

M. Ahmadieh Khanesar et al., *Sliding-Mode Fuzzy Controllers*, Studies in Systems, Decision and Control 357, https://doi.org/10.1007/978-3-030-69182-0_4

system dynamic, its order, nominal functions, and their parameter values, is not followed. Instead, the Lyapunov condition of stability in the form of $\dot{V} = S\dot{S} \leq -\eta S$ is required to be satisfied. The successful fulfillment of this inequality results in the finite-time convergence of the sliding surface, S, to zero, which in turn results in the stability of the system.

4.2 Fuzzy Logic System to Tune Sliding-Mode Controller Parameters

4.2.1 Boundary Layer with Constant Boundary Width

In this method, the sign function is replaced with an FLS. It is to be noted that chattering occurs when the system states are close to $s = 0$. In this solution, when the system states are far from the sliding manifold, the FLS acts in exactly the same way as the sign function. On the other hand, when the states are close enough to the sliding manifold using some sort of the IF–THEN rules, FLS eliminates the chattering phenomenon and make the closed-loop system less sensitive to noise. Hence, when using this type of controller, the benefits of the existence of a signum function in the SMC signal, which is to push the system states toward the sliding surface is maintained while chattering is prevented.

It is possible to use simple fuzzy mapping to replace the $sign$ function. In this case, two MFs used to fuzzify the sliding manifold are as follows:

$$\mu_P = \begin{cases} 0 & if \ s < -1 \\ \frac{s+1}{2} & if \ -1 < s < 1 \\ 1 & if \ s > 1 \end{cases} \qquad (4.1)$$

and

$$\mu_N = 1 - \mu_P. \qquad (4.2)$$

The fuzzy rules considered for such a system are defined as follows:

$$Rule\#1 \quad : \quad If \ s \ is \ N \ Then \ s_z = -1$$
$$Rule\#2 \quad : \quad If \ s \ is \ P \ Then \ s_z = 1.$$

The final output of the fuzzy system replacing the $sign$ function is inferred as follows:

$$s_z = \frac{\mu_P - \mu_N}{\mu_P + \mu_N}. \qquad (4.3)$$

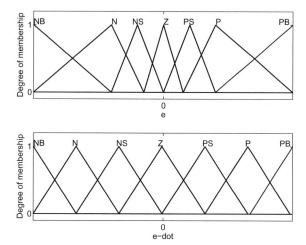

Fig. 4.1 Fuzzy MF considered for the sliding manifold (s)

This fuzzy system has been analyzed in detail in [6]. It was then concluded that the output of this fuzzy system is equal to that of a saturation function. Although this conclusion is correct, the main reason for such a simple result is that the considered MFs are very simple. It is possible to use more MFs to obtain a more smooth switching function, which acts different; than a saturation function does to obtain a superior performance.

It is also possible to obtain the fuzzy approximate of the *sign* function directly from the system states. In this case, the MFs act on position error and its time derivative (see Fig. 4.1). The rule base considered for this fuzzy system is summarized in Table 4.1, with N, Z, and P are selected as to be equal to -1, 0, and 1, respectively.

As can be seen from Fig. 4.2, this fuzzy system results in a more complex yet smooth function, which may replace the sign function. Furthermore, the sliding manifold in this case is nonlinear. As can be seen from Fig. 4.2b, although the slid-

Table 4.1 The fuzzy rule base for the more complex fuzzy system approximating the *sign* function

		e						
		NB	NM	NS	Z	PS	PM	PB
	NB	P	P	P	P	P	P	P
	NM	N	P	P	P	P	P	P
	NS	N	N	P	P	P	P	P
$\dot{e}$	Z	N	N	N	Z	P	P	P
	PS	N	N	N	N	N	P	P
	PM	N	N	N	N	N	N	P
	PB	N	N	N	N	N	N	N

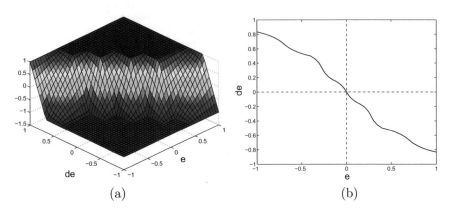

(a)   (b)

Fig. 4.2 a Surface of fuzzy system as a function of e and $\dot{e}$ replacing $sign$ function; **b** Nonlinear sliding manifold ($s = 0$)

ing manifold is nonlinear, it lies in the second and fourth quarters and therefore, introduces a stable dynamic for the system.

It is further possible to use a FLS for the coefficient of the $sign$ function. In order to obtain finite-time convergence of the sliding manifold to $zero$, the following equation must be satisfied.

$$s\dot{s} = -\eta|s|. \tag{4.4}$$

Let the system dynamics to be as follows:

$$\dot{x} = f(x) + g(x)u, \tag{4.5}$$

where $x \in R^n$, $f(x) : R^n \rightarrow R^n$, and $g(x) : R^n \rightarrow R^n$. Hence, the time derivative of the sliding manifold is obtained as follows:

$$\dot{s} = \frac{\partial s}{\partial x} f(x) + \frac{\partial s}{\partial x} g(x)u = -\eta sign(s). \tag{4.6}$$

The control signal to fulfill (4.6) is obtained as follows:

$$u = -\left(\frac{\partial s}{\partial x} g(x)\right)^{-1} \left(\frac{\partial s}{\partial x} f(x) + K sign(s)\right), \tag{4.7}$$

where $\eta \leq K$ and the existence of the inverse of $\frac{\partial s}{\partial x} g(x)$ is a necessary condition for the existence of the control signal.

Remark It is to be noted that if $K = \eta$, (4.6) is fulfilled. However, in order to be able to deal with unmodeled dynamics and possible external disturbances in the system, a larger value must be used for the parameter K.

Some observations can be made from the basic design procedure for a sliding-mode controller, which can be used to design a fuzzy rule-base supervisor which may improve the performance of the control loop and lessen undesirable effects such as chattering in the control signal. These observations are as follows [3]:

1. Since the desired behavior of the system is defined in terms of the sliding manifold and the system shows invariance properties on this surface, it is highly desirable to decrease the reaching time to the sliding manifold. In order to decrease the reaching time, it is possible to increase the parameter η. However, this parameter cannot be increased too much as a very large value for this parameter may result in saturation in the control signal. On the other hand, a very large value for η may result in chattering in the control signal.
2. It is, however, possible to use a small value for η, which results in small chattering.
3. The stability analysis done for the sliding-mode controller is based on the ideal *sign* function, which needs infinite switching frequency. However, realistic actuators suffer from delays and cannot fulfill the requirements imposed by the Lyapunov analysis to maintain the invariance property of the control system.

Based on these observations, it is suggested that when the system states are far from the sliding surface, a very high gain control signal acts on the system. On the other hand, if the system is close enough to the sliding manifold, the controller gain decreases to avoid crossing the sliding surface, which is the main reason for chattering. In this case, the following fuzzy rules are suggested to tune the controller gain online:

$$Rule\ 1 : \text{If } |s(x(t))| \text{ is SL Then } K(t) = KL$$
$$Rule\ 2 : \text{If } |s(x(t))| \text{ is SM Then } K(t) = KM$$
$$Rule\ 3 : \text{If } |s(x(t))| \text{ is SS Then } K(t) = KS$$
$$Rule\ 4 : \text{If } |s(x(t))| \text{ is SZ Then } K(t) = KZ$$

$$(4.8)$$

where SL, SM, SS, and SZ stand for large $|s|$, medium $|s|$, small $|s|$, and zero $|s|$, respectively. On the other hand, KL, KM, KS, and KZ also represent large K, medium K, small K, and zero K, respectively.

The overall K is obtained as follows:

$$K(t) = \frac{\mu_{SL} KL + \mu_{SM} KM + \mu_{SS} KS + \mu_{SZ} KZ}{\mu_{SL} + \mu_{SM} + \mu_{SS} + \mu_{SZ}}. \qquad (4.9)$$

Hence, the controller uses a large value for the parameter K, when s is far from *zero* and a small value for it when the parameter s is close to *zero*.

Fig. 4.3 Pendulum diagram

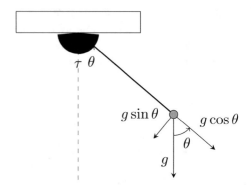

Example Consider a pendulum as presented in Fig. 4.3. The system dynamics are presented as follows:

$$\dot{x}_1 = x_2$$
$$\dot{x}_2 = \frac{g}{l}sin(x_1(t)) + \frac{1}{ml^2}u(t) \qquad (4.10)$$

with x_1 and x_2 being the system states, which represent θ and its time derivative, respectively. Furthermore, the parameters $m = 1$ Kg and $l = 1$ m represent the weight of the pendulum and its length, respectively. The parameter g represents the gravity acceleration, which is taken to be equal to $g = 9.8$ m/s^2.

The control objective is to track a desired angle θ_d. Hence, the following sliding manifold is designed.

$$s = \lambda e + \dot{e}, \qquad (4.11)$$

where e represents the tracking error and is defined as $e = \theta - \theta_d$. Hence, the equivalent control signal is obtained from $\dot{s} = 0$, as follows:

$$u_{eq} = ml^2\lambda(\dot{\theta} - \dot{\theta}_d) - ml^2\ddot{\theta}_d - mlgsin(\theta). \qquad (4.12)$$

It is supposed that the parameter g is uncertain, and it is estimated to be equal to $g = 9$ m/s^2. Hence, the overall control signal is obtained as follows:

$$u_{eq} = \lambda(\dot{\theta}_d - \dot{\theta}) + \ddot{\theta}_d - 9sin(\theta) - Ksign(s), \qquad (4.13)$$

where $K \geq 0.8$ to guarantee the convergence of the sliding manifold s to *zero*.

In the simulations, λ the sliding-mode parameter is selected as to be equal to 10, and the initial conditions for the system are selected as to be equal to $x_1 = 0$ and $x_2 = 0.2$ rad/s.

The MFs considered for the fuzzy system are selected as in Fig. 4.4. The consequent part parameters are selected as to be equal to $KZ = 0$, $KS = 0.4$, $KM = 0.8$, and $KL = 2$. On the other hand, in order to make a comparison, the constant parameter K in the conventional sliding-mode controller is selected as to be equal to $K = 0.8$. The comparison results between these controllers are illustrated in Fig. 4.4. As can be seen from Fig. 4.4, the fuzzy controller results in a stable response with less convergence time and less chattering in the control signal. Moreover, the evolution of the parameter K is presented in Fig. 4.5. As can be seen from Fig. 4.5, the parameter K has a larger value than the conventional method in the beginning, which results in faster convergence to the sliding surface; and finally, its value becomes smaller than its conventional counterpart, which results in smaller chattering in the control signal.

4.2.2 Boundary Layer with Adaptive Boundary Width

It is possible to use the fuzzy system to tune the width of *saturation* function which is normally used instead of the *sign* function in conventional approaches. It is known that a small boundary layer may result in chattering, while a large value for the width of this function may result in poor tracking performance of the system and cause steady state error in the system.

For example, the usage of adaptive boundary layer has been used for nonlinear hydraulic position control problem [1]. A similar method with different fuzzy system and rule base is investigated in [2]. In this case, the saturation function, which is used instead of *sign* function, is defined as follows:

$$sat(s) = \begin{cases} s/\phi & if \ |s/\phi| < 1 \\ sign(s) & if \ |s/\phi| \geq 1 \end{cases}, \tag{4.14}$$

where the fuzzy system is used to tune the parameter ϕ.

The method in [2] is based on the detection of chattering in the control signal. As mentioned earlier, chattering refers to high-frequency oscillations in the control signal. Hence, the time derivative of the control signal $\Gamma = |\dot{u}|$ can be used as a measure of the existence of chattering in the control signal (see Fig. 4.6).

Another variable, which can be used as the input of the fuzzy system, is the absolute value of the sliding variable (s). The fuzzy rule base, which is considered for this system, is presented in Table 4.2. The parameter ϕ is updated as follows:

$$\phi(t) = \phi(t-1) + \Delta\phi(t), \tag{4.15}$$

where the parameter $\Delta\phi$ is calculated using the fuzzy system. As can be seen from this table, the fuzzy rules are designed such that if there exists large chattering and the value of sliding variable is small, the value of the width of the boundary layer increases. On the other hand, when the sliding variable is "Big", the possibility of change in the sign of sliding variable is small and it is possible to decrease the

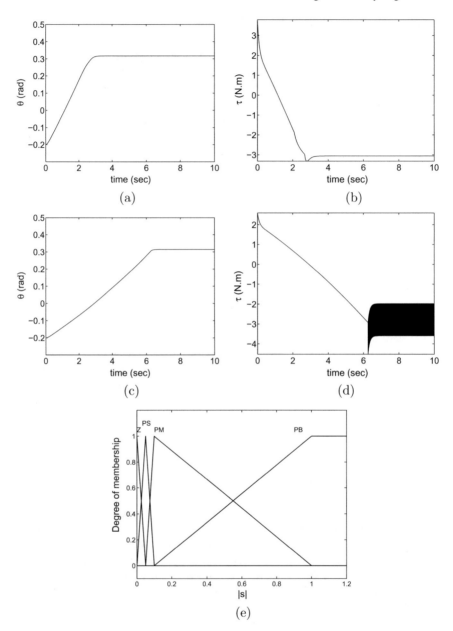

Fig. 4.4 **a** Response of the system when the fuzzy system is used **b** Control signal when the fuzzy system is used **c** Response of the system when the classical sliding-mode controller is used **d** Control signal when the classical sliding-mode controller is used **e** Fuzzy MF considered for the sliding manifold (s)

Fig. 4.5 The evolution of the parameter K when the fuzzy system is used

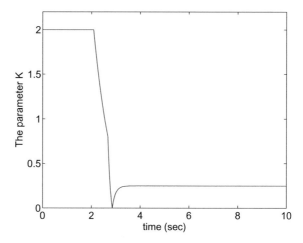

Fig. 4.6 The Chattering index based on $|\dot{u}|$ variable, which can be used as the input of the fuzzy system

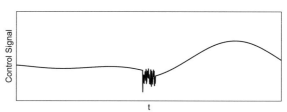

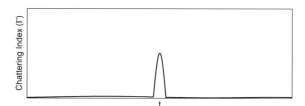

Table 4.2 The fuzzy rule base for the tuning of ϕ, the boundary layer width

		Γ	
		Small Γ	Big Γ
$|s|$	Big $|s|$	NB $\Delta\phi$	NS $\Delta\phi$
	Small $|s|$	0	PB $\Delta\phi$

parameter ϕ. When both the chattering index and the value of the sliding variable are small, it is possible to maintain the last value of the parameter ϕ.

It is to be noted that although adaptive tuning of the parameter ϕ may be more appropriate and decrease the chattering considerably, it may disturb the stability analysis. In other words, $\dot{\phi} \neq 0$ must be taken into account and stability analysis is done considering this fact [5].

4.3   Direct Sliding-Mode Fuzzy Logic Systems

In the direct sliding-mode fuzzy controller design approaches, a conventional SMC method is not designed. Instead, using the principles of SMC theory, a fuzzy controller is designed. In most of these approaches, the fuzzy rule base is designed such that $s\dot{s} \leq 0$.

Consider a second-order nonlinear dynamic system as follows:

$$\ddot{x}_1 = f(x) + g(x)u, \tag{4.16}$$

where $g(x) > 0$ and $f(x)$ are two unknown scalar functions of system states $x \in R^2$. The sliding manifold is taken as $s = \dot{x} + \lambda x$. Hence, the time derivative of the sliding surface is obtained as follows:

$$\dot{s} = f(x) + g(x)u + \lambda\dot{x}, \tag{4.17}$$

If a Lyapunov function is selected as follows:

$$V = \frac{1}{2}s^2 \tag{4.18}$$

its time derivative is obtained as follows:

$$\dot{V} = s\dot{s} = s(f(x) + g(x)u). \tag{4.19}$$

It is required from the Lyapunov theory that $s\dot{s} < 0$. Since $g(x) \geq 0$, in order to guarantee the condition imposed by the Lyapunov theory, the following two rules construct the basic idea for the design of a Lyapunov-based control system [7].

Rule 1 : If $s > 0$ and u is decreasing Then $s\dot{s}$ decreases
Rule 2 : If $s < 0$ and u is increasing Then $s\dot{s}$ decreases

$$. \tag{4.20}$$

Based on these principle rules, a fuzzy rule-based system is designed to determine the changes in the control signal Δu. In this case, the control signal is updated as

Table 4.3 Fuzzy rule base for the more complex fuzzy system approximating the *sign* function

$\dot{s}$ \ s	NB	NS	ZO	PS	PB
PB	ZO	NS	NB	NB	NB
PS	PS	ZO	NS	NB	NB
ZO	PB	PS	ZO	NS	NB
NS	PB	PB	PS	ZO	NS
NB	PB	PB	PB	PS	ZO

follows:

$$u(t+1) = u(t) + \Delta u(t), \tag{4.21}$$

where $\Delta u(t)$ is the output of a fuzzy system whose rule base is presented in Table 4.3.

This controller is implemented on a pneumatic servo system [7]. The controller they have designed resulted in *zero* steady-state error. Moreover, the controller works with a very little knowledge about the dynamic of the system and tunes the control signal independent of the plant model. Hence, this controller can easily be used to control other dynamic systems, provided that $g(x) > 0$.

Another model-free SMFC approach is designed using automata. In this case, if $s\dot{s} < 0$, then the previous change in the control signal is maintained. However, if $s\dot{s} > 0$, the previous change in the control signal is reversed. The rules used in this automata-based control system are as follows [8]:

If $sign(s\dot{s}) = -1$ and the previous control action was to increase Then keep on increasing
If $sign(s\dot{s}) = -1$ and the previous control action was to decrease Then keep on deceasing
If $sign(s\dot{s}) = 1$ and the previous control action was to increase Then decrease it
If $sign(s\dot{s}) = 1$ and the previous control action was to decrease Then increase it

In order to make sure that no chattering is visited, the following rules are considered.

If $sign(e(k)e(k-1)) < 0$ Then reduce $|\Delta u(k)|$
If $sign(e(k)e(k-1)) > 0$ Then maintain $|\Delta u(k)|$.

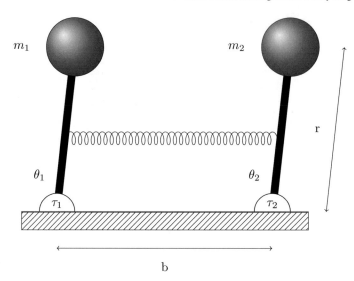

Fig. 4.7 Double-inverted pendulum system

Example In this part, a direct fuzzy logic control system is used to control a double inverted pendulum connected by a spring (See Fig. 4.7). Each pendulum is controlled by a torque-controlled servomotor at its base. The nonlinear dynamic model of this system is as follows:

$$\dot{x}_{11} = x_{12}$$
$$\dot{x}_{12} = \left(\frac{m_1 gr}{J_1} - \frac{kr^2}{4J_1}\right)sin(x_{11}) + \frac{kr}{2J_1}(l-b) + \frac{u_1}{J_1} + \frac{kr^2}{4J_1}sin(x_{21})$$
$$\dot{x}_{21} = x_{22}$$
$$\dot{x}_{22} = \left(\frac{m_2 gr}{J_2} - \frac{kr^2}{4J_2}\right)sin(x_{21}) - \frac{kr}{2J_2}(l-b) + \frac{u_2}{J_2} + \frac{kr^2}{4J_2}sin(x_{12}), \quad (4.22)$$

where $J_1 = 0.5\,\text{Kg}$, $J_2 = 0.625\,\text{Kg}$, $m_1 = 2\,\text{Kg}$, $m_2 = 2.5\,\text{Kg}$, $g = 9.81\,\text{m/s}^2$, $k = 100\,\text{N/m}$, $l = 0.5\,\text{m}$, $b = 0.4\,\text{m}$, and $r = 0.5\,\text{m}$.

Interval type-2 fuzzy logic controller is used to control this system, which is tuned based on the fact that to design a stable sliding-mode fuzzy controller, it is required that $\dot{V} = s\dot{s} \leq 0$. The tracking error signals in this system are defined as $e_1 = r_1 - x_{11}$ and $e_2 = r_2 - x_{21}$. In this case, the two sliding surfaces are defined as follows:

$$s_1 = \dot{e}_1 + \lambda_1 e_1$$
$$s_2 = \dot{e}_2 + \lambda_2 e_2.$$

$$(4.23)$$

The time derivative of the sliding surfaces (s_1) is obtained as follows:

$$\dot{s}_1 = \ddot{e}_1 + \lambda_1 \dot{e}_1$$
$$= \dot{r}_1 - \left(\frac{m_1 g r}{J_1} - \frac{kr^2}{4J_1} \right) sin(x_{11}) - \frac{kr}{2J_1}(l-b) - \frac{u_1}{J_1}$$
$$- \frac{kr^2}{4J_1} sin(x_{21}) + \lambda_1 \dot{e}_1. \tag{4.24}$$

Hence, when $s_1 < 0$, in order to make $s_1 \dot{s}_1$ negative, a negative value for the control signal u_1 must be selected. On the other hand, in the case of a positive value for s_1, it is required that a positive value for u_1 be selected. The inputs of the system are considered to be s_1 and $\dot{s}_1$. The MFs considered for the input values are considered to be as depicted in Fig. 4.8a. The rule base considered for the system is presented in Table 4.4. The type reducer used is "center-of-set type reducer without sorting requirement" [4]. Furthermore, there exists a constant term $-\frac{kr}{2J_1}(l-b)$ and

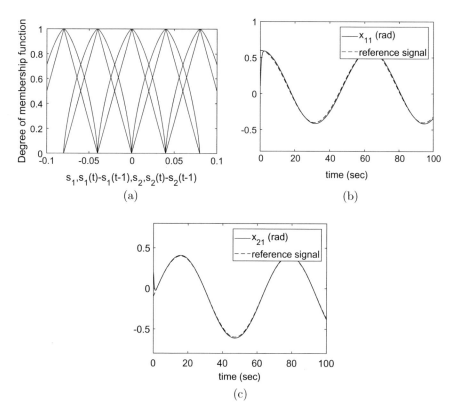

(a)

(b)

(c)

Fig. 4.8 a MFs considered for interval type-2 fuzzy logic control of double inverted pendulum **b** response of the double inverted pendulum x_{11} **c** The response of the double inverted pendulum x_{21}

Table 4.4 Fuzzy rule base for the more complex fuzzy system approximating the $sign$ function

		s				
		NB	N	Z	P	PB
	NB	PB	P	P	PS	PS
	N	P	P	PS	PS	PS
$\dot{s}$	Z	PS	PS	Z	NS	N
	P	NS	NS	NS	N	N
	PB	NS	NS	N	N	NB

$\frac{kr}{2J_2}(l-b)$ in the $\dot{s}_1$ and $\dot{s}_2$, respectively. This term is eliminated using the control signal. Hence, the output of the fuzzy system is added to the term which eliminates these constant terms. The reference signal r_1 is considered to be as follows:

$$r_1(t) = 0.1 + 0.5cos\left(\frac{t}{10}\right), \tag{4.25}$$

The reference signal r_2 is considered to be as follows:

$$r_2(t) = -0.1 + 0.5sin\left(\frac{t}{10}\right). \tag{4.26}$$

The tracking responses for this system are depicted in Fig. 4.8b, c.

References

1. Becan, M.R.: Fuzzy boundary layer solution to nonlinear hydraulic position control problem. Int. J. Mech. Aerosp. Ind. Mechatron. Manuf. Eng. **1**(5), 268–270 (2007)
2. Erbatur, K., Kawamura, A.: Chattering elimination via fuzzy boundary layer tuning. In: IECON 02, IEEE 2002 28th Annual Conference of the Industrial Electronics Society, vol. 3, pp. 2131–2136. IEEE (2002)
3. Ghalia, M.B., Alouani, A.T.: Sliding mode control synthesis using fuzzy logic. In: Proceedings of the 1995 American Control Conference, vol. 2, pp. 1528–1532. IEEE (1995)
4. Khanesar, M.A., Khakshour, A.J., Kaynak, O., Gao, H.: Improving the speed of center of sets type reduction in interval type-2 fuzzy systems by eliminating the need for sorting. IEEE Trans. Fuzzy Syst. **25**(5), 1193–1206 (2017)
5. Lee, H., Kim, E., Kang, H.-J., Park, M.: A new sliding-mode control with fuzzy boundary layer. Fuzzy Sets Syst. **120**(1), 135–143 (2001)
6. ODell, B.: Fuzzy sliding mode control: a critical review (1997)
7. Shin, M.-C., Ching-Sham, L.: Fuzzy sliding mode position control of a ball screw driven by pneumatic servomotor. Mechatronics **5**(4), 421–431 (1995)
8. Tzafestas, S.G., Rigatos, G.G.: Stability analysis of an adaptive fuzzy control system using petri nets and learning automata. Math. Comput. Simul. **51**(3), 315–339 (2000)

Chapter 5
Adaptive Sliding-Mode Fuzzy Control Systems: Gradient Descent Method

5.1 Introduction

GD is a computational optimization method, which is based on the first-order Taylor expansion of nonlinear functions. In order to find a local minimum for a nonlinear function, this algorithm uses the initial parameters of the nonlinear function and updates these parameters in the negative direction of the gradient of the function with respect to the parameters. This algorithm is one of the most frequently used algorithms. The benefit of this algorithm over intelligent optimization methods is that the latter are blind in that they do not consider the formula of the nonlinear function; however, the GD method uses the formula of the nonlinear function and wisely tries to minimize this function, which results in fast convergence, provided that the initial conditions for the parameters are selected appropriately and a suitable learning rate is selected.

In this chapter, the concept of GD is briefly discussed. The SMC theory-based cost function is defined and is minimized using the GD method. The cost function is defined such that it minimizes certain functions of sliding manifold to obtain the desired behavior of the system.

5.2 The Concept of the Gradient Descent Method

The GD method is used to optimize a nonlinear function. Let the unconstrained nonlinear minimization problem be as follows:

Minimize the scalar cost function

$F(x)$ subject to $x \in \mathbb{R}^n$.

© Springer Nature Switzerland AG 2021
M. Ahmadieh Khanesar et al., *Sliding-Mode Fuzzy Controllers*, Studies in Systems,
Decision and Control 357, https://doi.org/10.1007/978-3-030-69182-0_5

Since this cost function is a scalar function of its input vector x, in order to minimize this function, the gradient of the function must be determined. The gradient of the function $F(x)$ is represented by the symbol "∇F".

The Taylor expansion of $F(x(k+1))$ around $x(k)$ is given by

$$F(x(k+1)) = F(x(k)) + (\nabla F(x(k)))^T (x(k+1) - x(k)) + H.O.T. \quad (5.1)$$

The update rule for the parameter $x(k)$ is taken as follows:

$$x(k+1) = x(k) - \alpha \nabla F(x), \quad \forall \alpha > 0. \quad (5.2)$$

By applying the update rule of (5.2) in the Taylor expansion of (5.1), the following equation is obtained:

$$F(x(k+1)) = F(x(k)) - \alpha \| \nabla F(x(k))) \|^2 + H.O.T. \quad (5.3)$$

If higher-order terms in (5.3) are neglected, (5.3) implies that $F(x(k+1)) \leq F(x(k))$. The term $\alpha \| \nabla F(x(k))) \|^2$ dominates the higher-order terms near zero [1]. Hence, for small variations, $F(x(k+1)) < F(x(k))$. This is the main principle of GD. However, the signs and values of higher-order terms are undefined and may disturb the optimization problem if the function to be minimized is a highly nonlinear function. Moreover, the GD algorithm highly depends on the initial conditions and it may not result in the global minimization of a function.

5.2.1 Newton and Gauss–Newton Optimization Algorithm

As was mentioned earlier, GD is based on first-order Taylor expansion and takes its steps in the opposite direction of the gradient vector of the function with respect to its parameters. In order to obtain higher optimization performance, it is possible to take higher-order terms into the account. When second-order Taylor expansion is considered, the optimization algorithm is called Newton's optimization method. It is more complex than the GD method, but results in less iterations for the optimization process. This algorithm is also less sensitive to the initial conditions. The second-order derivative of the scalar function $F(x)$ with respect to the vector $x(k)$ is called Hessian. The second-order Taylor expansion of $F(x)$ around $x(k)$ is given as follows:

$$\begin{aligned} F(x) \approx G(x) = {}& F(x(k)) + (\nabla F(x(k)))^T (x - x(k)) \\ & + \frac{1}{2}(x - x(k))^T H(x(k))(x - x(k)), \end{aligned} \quad (5.4)$$

where $H(x)$ is the Hessian matrix of $F(x)$. Since (5.4) uses more terms of the Taylor expansion, it is more exact. In order to make the best possible step for the parameters, the gradient of $G(x)$ is used as follows:

$$\nabla G(x) = \nabla F(x(k)) + H(x(k))(x - x(k)) = 0. \tag{5.5}$$

The solution of which results in the following update rule for the parameters of the function:

$$x(k + 1) = -H^{-1}(x(k))\nabla F(x(k)). \tag{5.6}$$

Equation (5.6) is the Newton's optimization algorithm for $F(x)$. This update rule does not include a learning rate and design parameters that are difficult to select. In this case, the inverse of Hessian matrix acts as a learning rate for the GD method. What makes this algorithm difficult to implement and less interesting with respect to original GD method is that it includes the calculation of the Hessian matrix which must be inverted to obtain appropriate update value for the parameters of the function. The calculation of the inverse of a large scale matrix is a complex procedure whose complexity is as high as $O(n^3)$, in which, n is the number of the rows of the Hessian matrix to be inverted.

In order to add more degrees of freedom, the Newton's optimization algorithm (5.6) is modified by adding a learning as follows:

$$x(k + 1) = -\gamma (H(x(k)))^{-1}\nabla F(x(k)), \ 0 < \gamma < 1. \tag{5.7}$$

However, since the calculation of the Hessian matrix is complex as mentioned earlier, this matrix is approximated in order to reduce the complexity of the Newton's optimization method. Consider the cost function $F(x)$ to be in the following form:

$$F(x) = e^T(x)e(x), \tag{5.8}$$

where $e(x) \in \mathbb{R}^{m \times 1}$ is a vectorial function of x, which represents the difference between the desired value of the FNN and its output. The gradient of $F(x)$ is derived as follows:

$$\nabla F(x) = J^T(x)e(x), \tag{5.9}$$

in which J is the Jacobian matrix and is defined as follows:

$$J(x) = \begin{bmatrix} \frac{\partial e_1(x)}{\partial x_1} & \frac{\partial e_1(x)}{\partial x_2} & \cdots & \frac{\partial e_1(x)}{\partial x_n} \\ \frac{\partial e_2(x)}{\partial x_1} & \frac{\partial e_2(x)}{\partial x_2} & \cdots & \frac{\partial e_2(x)}{\partial x_n} \\ \vdots & \vdots & \ddots & \vdots \\ \frac{\partial e_m(x)}{\partial x_1} & \frac{\partial e_m(x)}{\partial x_2} & \cdots & \frac{\partial e_m(x)}{\partial x_n} \end{bmatrix}.$$

Furthermore, the Hessian matrix is obtained as follows:

$$\nabla^2 F(x) = J^T(x)J(x) + S(x), \qquad (5.10)$$

where

$$S(x) = \sum_{i=1}^{m} e_i(x)\nabla^2 e_i(x)$$

and $e_i(x)$ is the ith element of the vector $e(x)$. In order to reduce the complexity of optimization, $S(x)$ is neglected and Newton's update rules are approximated as follows:

$$x(k+1) = x(k) - [J^T(x(k))J(x(k))]^{-1}J^T(x(k))e(x(k)). \qquad (5.11)$$

The resulting update rule is called the Gauss–Newton optimization method.

5.2.2 Levenberg–Marquardt Optimization Algorithm

As can be seen from (5.11), in the last steps of optimization, it is possible that $J(x(k))J^T(x(k))$ may become singular or near singular. In order to solve this problem, the original update rule for the Gauss–Newton algorithm is modified as follows:

$$x(k+1) = x(k) - [J(x(k))J^T(x(k)) + \mu(k)I]^{-1}J(x(k))e(x(k)), \quad (5.12)$$

where I is an identity matrix of appropriate size and $\mu(k)$ has a positive value, which avoids the inversion of a singular matrix. It is to be noted that if a large value is selected for $\mu(k)$, (5.12) would be similar to a simple GD algorithm; on the other hand, if small values are selected for $\mu(k)$, (5.12) would converge to the Gauss–Newton algorithm of (5.11). Hence, in the first few steps of optimization, it is possible to use small values for $\mu(k)$ to obtain a Gauss–Newton algorithm, which is less sensitive to initial conditions; in the next steps, a larger value for $\mu(k)$ is selected to find the minimum of the nonlinear function. The flowchart of the Levenberg–Marquardt algorithm is illustrated in Fig. 5.1.

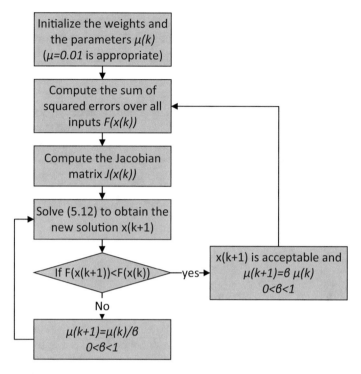

Fig. 5.1 The flowchart of LM algorithm

5.3   Sliding-Mode Theory-Based Cost Functions

The design of sliding-mode theory-based cost function is a very important step that may influence the overall performance of the system. The cost function to be minimized may include error signals, control signals, and time. If only an error signal is used, the performance of the system may be guaranteed while unnecessarily large control signals may be obtained. On the other hand, it is possible to include the control signal to obtain less control effort, which may result in decreased performance of the system.

In the design of a sliding-mode controller, in order to obtain a finite-time convergence, it is highly desired to consider the control signal such that $\dot{s} + Dsign(s) = 0$. Hence, it is possible to design the cost function based on this parameter as follows:

$$J = \frac{1}{2}\left(\dot{s} + \eta sign(s)\right)^2. \tag{5.13}$$

If this cost function is minimized, the dynamic equation governing the system becomes $\dot{s} = -\eta sign(s)$, which guarantees finite-time convergence of the sliding manifold to *zero*. On the other hand, it is possible to design the cost function to avoid the *signum* function. Such a cost function is designed as follows:

$$J = \frac{1}{2}(\dot{s} + Ds)^2.$$ (5.14)

Although this cost function does not include any signum function, it does not provide finite-time convergence of the sliding manifold to *zero*. The dynamic equation governing the system when this cost function converges to zero is obtained as $\dot{s} = -Ds$, which is a stable behavior and results in the system being stabilized.

On the other hand, if the control effort is important due to less energy to be consumed or possible saturation in the input of the system, the control signal may need to be included in the cost function. In this case, the cost function includes the square of the control signal as follows:

$$J = \frac{1}{2}(\dot{s} + Ds)^2 + \lambda u^2,$$ (5.15)

where λ is a design parameter that has a high impact on the performance of the system. If a small value for λ is used, higher performance for the system is obtained. However, if a large value for the parameter λ is used, a small value for the control signal is obtained while the performance of the system may be lost. Hence, the selection of the parameter λ is a trade-off between high performance for the system and the magnitude of the control signal, which must be tuned carefully by the designer.

Another phenomenon that is important is the existence of chattering in the control signal. The changes in the control signal $\dot{u}(t)$ is a measure of chattering and may be included in the cost function to avoid this undesirable phenomenon.

5.4   Gradient Descent-Based Sliding-Mode Fuzzy Control of a DC–DC Converter

In this part, an IT2NFC is used to control a Buck DC–DC converter. The IT2FLS is optimized using GD learning algorithm. The adaptive controller makes it possible to control the system with higher performance in terms of the rise time of the system and smoother step response. In addition, the use of IT2MFs makes it possible to deal with uncertainties in terms of noise more effectively. It is shown that the proposed method can control the DC–DC converter in the presence of high levels of noise. The simulations are done in the presence of different levels of noise. It is shown that as more power of noise is injected into the system, the type-2 controller outperforms its type-1 counterpart with a higher percentage.

Power electronic converters are used extensively in personal computers, mobile devices, and adapters to provide the required DC voltage levels. A stable and smooth DC voltage level is vital in many electronic devices. A slight deviation of the output DC voltage from its desired values may cause serious damages in most high-frequency electronic devices. Since obtaining analog control signal in power electronic applications is difficult (if not impossible) in most cases, different switching control methods have been proposed. Among these switching methods, pulse width modulation (PWM) is the most frequently used approach. Several control approaches are investigated to control the width of PWM in DC–DC converters. For instance, PI controllers, the sliding-mode control method, and pole placement are used to control a PWM DC–DC converter. However, most of these control approaches are highly dependent upon a sufficiently exact model of the system and its dynamics. Moreover, tuning of their parameters needs a set of trial-and-error experiments [2, 3].

In recent years, there has been a tremendous amount of activity toward the development of efficient control strategies to increase the performance of DC–DC converters by using fuzzy logic controller, neural networks, and fuzzy neural network controller. For instance, in [4], a fuzzy logic controlled (FLC) buck–boost DC–DC converter for solar energy-battery systems has been studied. In [5], a fuzzy PID controller is designed and applied to a DC–DC converter. In [6], an adaptive fuzzy neural network control scheme is designed for the voltage tracking control of a conventional DC–DC boost converter. Neural network control of DC–DC converters is also studied in [7]. Type-2 FLC has also been previously used to control a DC–DC converter [2, 3]. The motivation of this paper is to design a IT2FLS for the DC–DC converter. To the best knowledge of the author of this article, IT2FLS has not been previously applied to control the DC–DC converter. The parameters of the consequent part of T2FNC are tuned using a GD training algorithm. Using IT2FLS, it is expected to control the DC–DC converter with higher performance in terms of rise time and settling time. Moreover, since type-2 MFs are used in the controller, the system performs better in the presence of measurement noise in the system. Therefore, this paper is another example of the superior performance of IT2FLS over type-1 counterpart in the presence of high levels of measurement noise.

5.4.1 The Model of DC–DC Converter and its Computer Simulation

The circuit diagram and the structure of DC–DC converter are shown in Fig. 5.3. The basic operation of the buck DC–DC converter can be explained as follows. As can be seen from Fig. 5.3, the circuit consists of an inductor and two semiconductor switches (usually a transistor and a diode) that control the inductor. When the switch (S) is closed, the inductor is connected to the DC voltage source, and the inductor stores energy. But when S is opened, the diode (D) is closed, and the stored energy of the diode is discharged into the load. In order to have more realistic simulations, the resistors r_L and r_C are considered in the series with the inductor and the capacitor (Fig. 5.2).

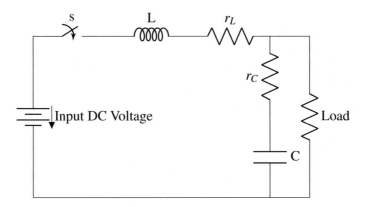

Fig. 5.2 The DC–DC converter circuit

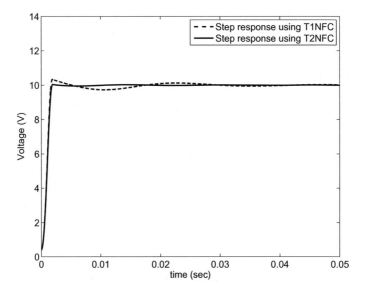

Fig. 5.3 The comparison of step responses of IT2FLS and T1NFC

The state equation for a buck DC–DC converter can be written as follows [3]:

$$
\begin{aligned}
\dot{x} &= A_i x + B_i V_s, \, i = 1, 2 \\
y &= C_i x
\end{aligned}
\tag{5.16}
$$

where $x = [i_L v_C]^T$, and A_i, B_i, and C_i are system matrices of the constituent linear circuits, and V_s is the input DC voltage of the converter. The DC–DC converter operates in two modes. When the semiconductor switch is on, the states matrices with subscript "1" represent the dynamical model of the system, while those with subscript "2" represents the dynamical model of the system when the semiconductor switch is off. The system matrices can be obtained for different operating modes as follows:

$$A_1 = A_2 = \begin{bmatrix} -\frac{1}{L}\left(\frac{Rr_C}{R+r_C} + r_L\right) & -\frac{1}{L}\left(\frac{R}{R+r_C}\right) \\ \frac{1}{C}\left(\frac{R}{R+r_C} + r_L\right) & -\frac{1}{C}\left(\frac{1}{R+r_C}\right) \end{bmatrix}, \tag{5.17}$$

$$B_1 = \begin{bmatrix} \frac{1}{L} \\ 0, \end{bmatrix} \tag{5.18}$$

$$B_2 = \begin{bmatrix} 0 \\ 0, \end{bmatrix} \tag{5.19}$$

$$C_1 = C_2 = \begin{bmatrix} \frac{Rr_C}{R+r_C} & \frac{R}{R+r_C} \end{bmatrix}. \tag{5.20}$$

The state-space averaging method is a very useful technique to analyze the low-frequency small-signal performance of switch circuits [3]. It is applicable when the converter switching period is short as compared to the response time of the output voltage. Using the state-space averaging method, the state equation can be obtained as follows:

$$\dot{x} = Ax + BV_s$$
$$y = V_o = Cx, \tag{5.21}$$

in which V_o is the output DC voltage of the converter and $A = dA_1 + (1 - d)A_2$, $B = dB_1 + (1 - d)B_2$, and $C = dC_1 + (1 - d)C_2$, and d is the duty cycle of the switching. The control objective is to control the duty cycle so that the output voltage can supply a smooth voltage in the presence of disturbances in terms of ripple and changes of the input DC voltage and variations of the output current of the system. In this paper, in order to achieve more accurate and realistic simulations, the DC–DC converter is simulated in the Simulink®/Power system blockset is used. The basic circuit diagram of the DC–DC converter is drawn in Simulink. The use of Simulink software makes it possible to use embedded functions and/or S-functions to implement a wide class of controllers as required by the designer. The values of the parameters of the DC–DC converter are considered as $L = 150\,\mu H$ and $C = 1\,mF$ and the frequency of PWM is considered to be 50 KHz.

The type-2 fuzzy structure used is A2-C0 fuzzy system with its type reducer selected to be Nie–Tan type reducer which is described in Chap. 2. Among the various IT2MFs mentioned in the literature, namely triangular, Gaussian, trapezoidal, sigmoidal, pi-shaped, etc., the elliptic IT2MF is used. This membership function has certain values on both ends of the support and the kernel and some uncertain values on the other parts of the support. The mathematical expression for the membership function is expressed as follows:

$$\tilde{\mu}(x) = \begin{cases} \left(1 - \left|\frac{x-c}{d}\right|^a\right)^{\frac{1}{a}}, & if |x - c| \le d \\ 0 & else, \end{cases} \tag{5.22}$$

where c and d are the center and the width of the membership function and x is the input value to the membership function. The parameters a_1 and a_2 determine the width of the uncertainty of the elliptic membership function to be selected as follows:

$$a_2 \le 1 \le a_1. \tag{5.23}$$

These parameters can be selected as some constants, or they can be tuned adaptively.

The parameters $x_1, x_2, \ldots, x_n$ are the input variables, $u_j (j = 1, \ldots, M)$ are the output variables, and $\tilde{A}_{ij}$ is an IT2MFs for jth rule and the ith input. α_{ij} and $\beta_j (i = 1, \ldots, n, j = 1, \ldots, M)$ are the parameters in the consequent part of the rules. The output of the fuzzy system in closed form is achieved by [15]

$$Y_{TSK1} = \frac{\sum_{j=1}^{M} \underline{w}^j F_j}{\sum_{j=1}^{M} \underline{w}^j + \sum_{j=1}^{M} \overline{w}^j} + \frac{\sum_{j=1}^{M} \overline{w}^j F_j}{\sum_{j=1}^{M} \underline{w}^j + \sum_{j=1}^{M} \overline{w}^j}, \tag{5.24}$$

where $\overline{w}^j$ and $\underline{w}^j$ are given by

$$\begin{aligned} \underline{w}^j(x) &= \underline{\mu}_{\tilde{F}_1^j}(x_1) * \cdots * \underline{\mu}_{\tilde{F}_n^j}(x_n) \\ \overline{w}^j(x) &= \overline{\mu}_{\tilde{F}_1^j}(x_1) * \cdots * \overline{\mu}_{\tilde{F}_n^j}(x_n), \end{aligned} \tag{5.25}$$

$$Y_{TSK1} = \frac{\sum_{j=1}^{M} (\underline{w}^j + \overline{w}^j) F_j}{\sum_{j=1}^{M} \underline{w}^j + \sum_{j=1}^{M} \overline{w}^j}. \tag{5.26}$$

In this way, the firing of each rule is defined as follows:

$$r_j = \frac{\underline{w}^j + \overline{w}^j}{\sum_{j=1}^{M} \underline{w}^j + \sum_{j=1}^{M} \overline{w}^j}. \tag{5.27}$$

and

$$F_j = \sum_{i=1}^{n} \alpha_{ij} x_i + \beta_j$$

5.4.2 Design of IT2FLS for DC–DC Converter

5.4.2.1 Parameter Update Rules

The design of T2FLS includes the determination of the unknown parameters that are the parameters of the antecedent and the consequent parts of the fuzzy IF–THEN rules. The cost function to be optimized includes the sliding-mode parameter and is defined as follows:

$$J = \frac{1}{2}s^2, \tag{5.28}$$

where $s = \lambda e + \Delta e$ and e, and the error signal between the reference input and the output voltage of DC–DC converter is defined as follows:

$$e(t) = V_d(t) - V(t), \tag{5.29}$$

where $V_d(t)$ and $V(t)$ are the desired and the current voltage outputs of the DC–DC converter, respectively. The parameters α_{ij}'s, β_j's, a_{1ij}'s, and a_{2ij}'s are among the adjustable parameters using the GD method, which are optimized as follows:

$$\alpha_{ij}(t + 1) = \alpha_{ij}(t) - \gamma \frac{\partial J}{\partial \alpha_{ij}}, \tag{5.30}$$

$$a_{1ij}(t + 1) = a_{1ij}(t) - \gamma \frac{\partial J}{\partial a_{1ij}}, \tag{5.31}$$

$$a_{2ij}(t + 1) = a_{2ij}(t) - \gamma \frac{\partial J}{\partial a_{2ij}}, \tag{5.32}$$

where γ is the learning rate. The derivatives in (5.30) are calculated as follows:

$$\frac{\partial J}{\partial \alpha_{ij}} = \frac{\partial J}{\partial u} \frac{\partial u}{\partial F_j} \frac{\partial F_j}{\partial \alpha_{ij}}. \tag{5.33}$$

The derivatives in (5.31)–(5.32) are determined by the following formulas:

$$\frac{\partial J}{\partial a_{1ij}} = \sum_j \frac{\partial J}{\partial u} \frac{\partial u}{\partial \bar{w}_j} \frac{\partial \bar{w}_j}{\partial \bar{\mu}_{ij}} \frac{\partial \bar{\mu}_{ij}}{\partial a_{1ij}}, \tag{5.34}$$

$$\frac{\partial J}{\partial a_{2ij}} = \sum_j \frac{\partial J}{\partial u} \frac{\partial u}{\partial \underline{w}_j} \frac{\partial \underline{w}_j}{\partial \underline{\mu}_{ij}} \frac{\partial \underline{\mu}_{ij}}{\partial a_{2ij}}, \tag{5.35}$$

The parameters of the T2FLS can thus be updated using (5.30)–(5.32) together with (5.34)–(5.35).

The derivatives in (5.33) are calculated as

$$\frac{\partial J}{\partial u} = -Ja.\left(V_d(t) - V(t)\right), \tag{5.36}$$

where Ja is the Jacobian of the DC–DC converter. The Jacobian of the system can be found using the model of the system as in (5.16). As can be seen from this equation, the Jacobian of the system has a constant value equal to the input gain of the system.

$$\frac{\partial u}{\partial \beta_j} = 1, \tag{5.37}$$

$$\frac{\partial u_j}{\partial \alpha_{ij}} = r_j x_i. \tag{5.38}$$

5.4.3 Simulation Results

In this part, the simulation results of applying IT2FLS to the DC–DC converter are gathered. The GD update rule, as summarized in the previous section, is applied to control the system. As mentioned earlier, GD-based learning algorithm is applied to the parameters of the premise and the consequent part of IT2FLS. The sum of the square of the error of voltage of the DC–DC converter is the cost function used in the simulation studies.

Figure 5.3 shows the comparison of the output voltage of the DC–DC converter for T1NFC and T2NFC. As can be seen from the figure, the response of applying type-2 controller is less oscillating. The rise time and settling time of the system are

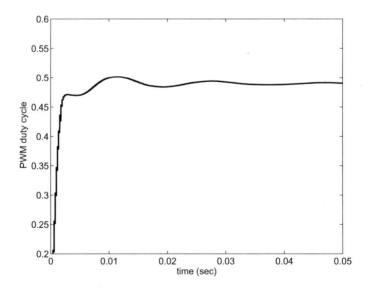

Fig. 5.4 The adaptive PWM duty cycle produced by T2NFC

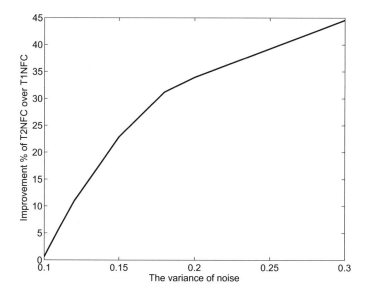

Fig. 5.5 The comparison of percentage of improvement of IT2FLS over T1NFC

less than 30 ms, which is much better than the corresponding values reported in [2, 3], in which, a non-adaptive T2FS is used to control the DC–DC converter.

Figure 5.4 shows the duty cycle of PWM to switch the switch used in the structure. The figure shows that the adaptation law completely tunes the duty cycle.

A comparison between IT2FLS and T1NFC is performed, in which, the Gaussian noise is added to the measurement in both type-1 and type-2 cases. Ten different simulation studies are carried out for different variances of the noise. The results can be seen in Fig. 5.5. From the figure, it can be observed that as more noise power is injected into the system, the IT2FLS outperforms its type-1 counterpart with a higher percentage. This is quite expected from an IT2FNC that when more power of the noise is injected into the system, the percentage of improvement of performance increases. This can be viewed as another example in which T2FSs lead to higher performance in the system.

5.5   Application to Control an IM

The main reasons for the widespread use of IMs in the industry are that they can be used with variable speed, they are low-cost, they can produce high torques, and they benefit from high reliability. Moreover, they use AC power, which eliminates the need for converters. The dynamic model of IMs is complex and nonlinear. One of the most frequent control methods to control an IM is field-oriented control, which results in high-performance system response. In this method, the nonlinear dynamics of the motor is simplified to dynamic equations governing a separately excited DC motor. The benefit of this control method is that it can also be used in sensorless control of IM because it needs the feedback from the flux and the state variable of IM. In this section, using a cost function that includes sliding mode, an NFCS for the speed control of an IM is presented.

5.5.1 Field-Oriented Control of IM

5.5.1.1 Dynamic Model of IM

The mathematical model of an IM is expressed in (5.39), which shows that the dynamical model of an IM is nonlinear [8, 9] :

$$\dot{\lambda}_{dr} = -\frac{1}{\tau_r}\lambda_{dr} + \omega_m\lambda_{qr} + \frac{L_m}{\tau_r}I_{ds}$$

$$\dot{\lambda}_{qr} = -\frac{1}{\tau_r}\lambda_{qr} - \omega_m\lambda_{dr} + \frac{L_m}{\tau_r}I_{qs}$$

$$\dot{I}_{ds} = \frac{\alpha}{\tau_r L_1}\lambda_{dr} + \frac{\alpha}{L_1}\omega_m\lambda_{qr} - \frac{1}{\tau_1}I_{ds} + \omega_s I_{qs} + \frac{1}{L_1}V_{ds}$$

$$\dot{I}_{qs} = \frac{\alpha}{\tau_r L_1}\lambda_{qr} - \frac{\alpha}{L_1}\omega_m\lambda_{dr} - \frac{1}{\tau_1}I_{qs} - \omega_s I_{ds} + \frac{1}{L_1}V_{qs}$$

$$\dot{\omega}_m = \frac{1}{J}\left[\frac{L_m}{L_m + L_r}\left(\lambda_{dr}I_{qs} - \lambda_{qr}I_{ds}\right) - T_L - B\omega_m\right], \qquad (5.39)$$

where the subscripts s and r refer to the stator and rotor, respectively, and subscript d and q denote the mathematical model in a synchronous rotating reference frame for a three-phase IM. The dots over each variable imply its time derivative, and Table 5.1 shows the name of parameter in the dynamic model of IM.

The following equation presents the relationship between the electromagnetic torque of the motor (T_e) and the radial speed of the motor ω_m:

$$T_e = J\frac{d\omega_m}{dt} + B\omega_m + T_L, \qquad (5.40)$$

where B is the viscous friction.

5.5.1.2 The Structure of Field-Oriented Control

In general, torque and speed control of a three-phase IM is more difficult than that of a DC machine. This is mainly because of the interactions between the stator and rotor field whose orientation is not held spatially at 90° but vary with operating condition.

Table 5.1 Parameter of dynamic model of IM

Parameters	Definition
λ_{dr}	Flux linkage of rotor in d axis
λ_{qr}	Flux linkage of rotor in q axis
I_{ds}	Stator current in d axis
I_{qs}	Stator current in q axis
ω_m	Motor speed
J	Inertia
L_s	Self-inductance of stator
L_r	Self-inductance of rotor
L_m	Mutual inductance
V_{qs}	Stator voltage in q axis
V_{ds}	Stator voltage in d axis
R_s	Stator resistance
R_r	Rotor resistance
ω_s	Speed of the stator magnetic field
T_L	Load torque
τ_1	Time constant
$\tau_r = \frac{L_r}{R_r}$	Rotor time constant
$\alpha = \frac{L_m}{L_r}$	By definition
$L_1 = L_s - \alpha L_m$	By definition
$R_1 = R_s + \alpha^2 R_r$	By definition
B	Viscous friction

If a synchronously rotating qd frame is selected, in which the d axis is precisely adjusted with the rotor field, the q component of the rotor flux (λ_{qr}) in the chosen reference frame would be zero, where λ_{qr} is flux rotor.

It is considered that

$$\lambda_{qr} = 0, \quad \dot{\lambda}_{qr} = 0; \tag{5.41}$$

Using (5.41), the desired rotor flux linkage in term of I_{ds} can be found from (5.39) as follows:

$$\lambda_{dr} = \frac{R_r L_m}{R_r + L_r p} I_{ds}, \tag{5.42}$$

where the operator p represents the time derivative operator. According to (5.39), the slip angular velocity can be estimated using λ_{dr}, as shown in (5.42), and I_{qs} is as follows:

$$\omega_s - \omega_m = \frac{R_r L_m}{L_r} \frac{I_{qs}}{\lambda_{dr}}. \tag{5.43}$$

The estimated slip angular velocity is obtained as follows:

$$\lambda_{dr} = L_m I_{ds}$$
$$\omega_s - \omega_m = \frac{R_r}{L_r} \frac{I_{qs}}{I_{ds}}. \tag{5.44}$$

With the implementation of direct field-oriented control, the electromagnetic torque can be shown as follows:

$$T_e = -\frac{3}{2} \frac{P}{2} \frac{L_m}{L_r} \lambda_{dr} I_{qs}. \tag{5.45}$$

5.5.1.3 Direct Field-Oriented Control Using Interval Type-2 Fuzzy Neural Network Structure

To deal with the control of a three-phase IM in a similar way to controlling a DC motor, it is required that the measured a-b-c stator current be transformed to the stationary q-d current as follows:

$$I_{qs} = \frac{2}{3} I_{as} - \frac{1}{3} I_{bs} - \frac{1}{3} I_{cs}$$
$$I_{ds} = \frac{1}{\sqrt{3}} (I_{cs} - I_{bs}). \tag{5.46}$$

Considering the following equations:

$$\lambda_m = L_m (I_{qs} + I_{qr})$$
$$\lambda_{qr} = \frac{L_r}{L_m} \lambda_{mq} - L_{lr} I_{qs}$$
$$\lambda_{dr} = \frac{L_r}{L_m} \lambda_{md} - L_{lr} I_{ds}, \tag{5.47}$$

where λ_{mq} and λ_{md} are the flux resultant measured in the q and d axis, respectively, the magnitude of the flux rotor and angle ρ are determined as follows:

$$\sin\left(\frac{\pi}{2} - \rho\right) = \cos\rho = \frac{\lambda_{dr}}{|\lambda_r|}$$
$$\cos\left(\frac{\pi}{2} - \rho\right) = \sin\rho = \frac{\lambda_{qr}}{|\lambda_r|}$$
$$|\lambda_r| = \sqrt{\lambda_{qr}^2 + \lambda_{dr}^2}. \tag{5.48}$$

According to the electromagnetic torque equation based on the direct field-oriented method,

$$T_e = K_t I_{qs}^*$$

$$K_t = \frac{3P}{2} \frac{L_m^2}{L_r} I_{ds}^*. \tag{5.49}$$

The proposed controller tunes electromagnetic torque by using an appropriate I_{qs}^* (torque current command), while the rotor flux is fixed by the adjustment of I_{ds}^* (flux current command). Substituting (5.49) and the fixed value of the rotor flux, which is controlled by the proposed controller, in the dynamic equation of the motor speed, the final equation for the speed control of IM is obtained as follows:

$$J \frac{dw(t)}{dt} = \frac{3\pi}{4} \cdot \frac{L_m^2}{L_r} \cdot I_{ds}^* \cdot I_{qs}^* - B \cdot w(t) - T_L. \tag{5.50}$$

5.5.2 Interval Type-2 Fuzzy Neural Network Controller

The fuzzy control of IMs is considered in several works. In particular, the sliding-mode fuzzy controller has found its position in different control methods used for controlling an IM. For instance, this type of controller has been used in a field-programmable gate array to control the mover position of a linear IM drive. Uncertainties in terms of friction have been compensated for using this method [10].

A combination of a sliding-mode controller with a PI-fuzzy logic-based can be used to control IMs. This combination makes it possible to benefit from the sliding-mode controller for the transient response and PI-like fuzzy controller for the steady state [11]. For more existing researches on the control of IMs including sliding-mode fuzzy controllers, different surveys including [12–14] can be referenced.

The IT2FLS developed for IMs in this section benefits from elliptic type-2 MFs and BMM approximate defuzzifier. The objective function used in this case includes sliding manifold to adjust the parameters of the controller in the presence of uncertainties, disturbances, and noise [15]. The main reason behind the usage of the type-2 fuzzy-based controller is that IMs suffer from time-varying parameters such as the values of resistances of the IM coils due to high current. Moreover, the dynamic equations considered for these structures are simplified to be handled successfully. Hence, since the new studies on IT2FSs show that they can handle uncertainties and noise, they are preferred to be used in this case.

The sliding surface considered is as follows:

$$S = \dot{e} + \lambda e, \tag{5.51}$$

where $e = w - w^*$ is the speed error and $\dot{e}$ is its time derivative. This equation is used in the adaptive mechanism in the structure of the proposed controller. The control strategy that is used here guarantees that the trajectories of the system (e and $\dot{e}$) move toward the sliding surface $(S=0)$ from any initial conditions and stay on it if the following condition is satisfied.

The IT2FLS considered as the controller in this case benefits from elliptic IT2MFs whose right-most and left-most values, which are expressed by $\overline{\mu_i}$ and $\underline{\mu_i}$, respectively, are defined as follows:

$$\overline{\mu}_i = \left(1 - \left(\left|\frac{x - c}{d}\right|\right)^{a_1}\right)^{\frac{1}{a_1}}$$

$$\underline{\mu}_i = \left(1 - \left(\left|\frac{x - c}{d}\right|\right)^{a_2}\right)^{\frac{1}{a_2}}. \tag{5.52}$$

The firing of the rules of IT2FS is obtained using a t-norm operator that implements the logic "AND" operator as follows:

$$\overline{w}_i = \overline{\mu}_1^i \times \overline{\mu}_2^i \times \cdots \times \overline{\mu}_n^i$$

$$\underline{w}_i = \underline{\mu}_1^i \times \underline{\mu}_2^i \times \cdots \times \underline{\mu}_n^i, \tag{5.53}$$

where $\overline{w}_i$ and $\underline{w}_i$ denote the upper and the lower outputs for the ith node in layer three and "n" is the number of membership functions.

The final output of the IT2FLS is calculated as follows:

$$Y = \frac{q \sum_{i=1}^{M} \underline{w}_i f_i}{\sum_{i=1}^{M} \underline{w}_i} + \frac{(1 - q) \sum_{i=1}^{M} \overline{w}_i f_i}{\sum_{i=1}^{M} \overline{w}_i}, \tag{5.54}$$

where Y is final output, and f_i is the value in consequent part as follows:

$$\theta_i = a_0^i + a_1^i x_1 + a_2^i x_2 + \cdots + a_n^i x_n. \tag{5.55}$$

In the above equation, $(x_1 \ldots x_n)$ are the inputs, n is the number of inputs, i is the rule number, and $(a_0 \ldots a_n)$ are constant values.

As was mentioned earlier, in order to obtain the desired performance for the system, it is required that the sliding manifold, which is considered as in (5.51), be equal to *zero*. In order to fulfill this goal, the cost function considered for the system is as follows:

$$J = (\dot{S} + DS)^2. \tag{5.56}$$

If this cost function is equal to *zero*, it follows that

$$\dot{S} = -DS, \tag{5.57}$$

which guarantees the infinite-time convergence of S to *zero* at the rate of D.

The control signal (I_{qs}) used for the speed control of IM based on field-oriented control is calculated by the proposed controller and the adaptive mechanism based on the value of the sliding surface S, and its derivative tunes the parameters of the controller. The adaptive mechanism in this structure uses GD algorithm for updating the parameters of the proposed controller.

The parameter update rule considered for updating θ_i is as follows:

$$\theta_i^{new} = \theta_i + \gamma_1 . T_S . \left[\frac{\underline{w_i}}{\sum \underline{w_i}} \cdot q + \frac{\overline{w_i}}{\sum \overline{w_i}} \cdot (1-q) \right] \times \left(\dot{S} + DS \right), \quad (5.58)$$

where T_S is the sampling time. Similarly, the GD-based parameter update rules considered for q are as follows:

$$q^{new} = q + \gamma_1 . T_S . \left[\frac{\sum f_i \underline{w_i}}{\sum \underline{w_i}} \cdot q - \frac{\sum f_i \overline{w_i}}{\sum \overline{w_i}} \right] \times \left(\dot{S} + DS \right). \quad (5.59)$$

It is to be noted that $0 \le q \le 1$ tunes the sharing of the right-most and left-most values of the output of IT2FLS.

5.5.2.1 Results

Seven MFs are considered for input e, and the total number of MFs considered for its time derivative is equal to 5. The θ_is, which are the consequent part parameters, are initialized by $zero$, and the sliding surface is selected as follows:

$$s = \dot{e} + \lambda e, \quad (5.60)$$

where λ is selected as being equal to 5 and the sample time considered for the simulations is selected as $T_s = 1$ ms. Table 5.2 summarizes the IM parameter values [16]. The parameter update rules used are as in (5.58) and (5.59).

Table 5.2 Value parameter of motor

Parameters	Values
R_r	$0.045\,\Omega$
R_s	$0.045\,\Omega$
L_r	$1.927\,H$
L_s	$1.927\,H$
L_m	$1.85\,H$
J	$58.5\,kg.m^2$
$I_{ds\,max}$	$50\,A$
T_L	$50\,N.m$

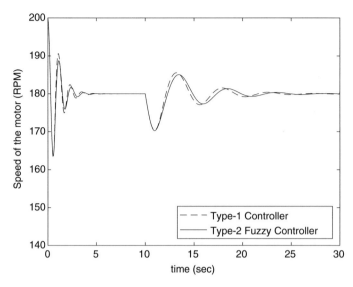

Fig. 5.6 The response of the closed-loop control system controlling IM in the presence of a change in the parameter

As can be seen from Table 5.2, the maximum output current of the inverter is equal to $50A$. Initially, it is assumed that the speed of the motor is equal to 200 and the reference signal is selected as being equal to 180. Furthermore, in order to investigate the effects of change in the parameters, it is assumed that the parameters of the IM may suddenly vary. It is assumed that the value of L_m suddenly changes to $0.4625\,H$ at $t = 10\,s$. The simulation results are depicted in Fig. 5.6. As can be seen from the figure, the type-2 system has a superior performance in the presence of parameter change as it results in a response with slightly less overshoot values. The effect of changes in load is investigated in Fig. 5.7. In this test, the load is changed from $T_L = 1000\,\text{Nm}$ to $T_L = 100\,\text{N.m}$ at $t = 10\,s$ and to $T_L = 3000\,\text{Nm}$ at $t = 15\,s$; as can be seen from this figure, a slightly superior performance for IT2FLCs with respect to T1FCs can be observed.

Finally, the tracking response of the system when the reference signal varies between 180 and 160 is illustrated in Fig. 5.8. As can be seen from the the figure, tracking is obtained with a high performance in both type-1 and type-2 fuzzy logic cases.

The result obtained during the simulations of the two controllers shows that as time goes, the overshoot in the response of the system decreases. This characteristic is due to the controller learning feature.

In summary, the results show the learning capability of T1FLC and IT2FLC when they are used to control an IM. It is further observed that IT2FLC demonstrates slightly superior performance with respect to its type-1 counterpart in terms of overshoot, damping, and response to changes in load and parameter. Hence, it is

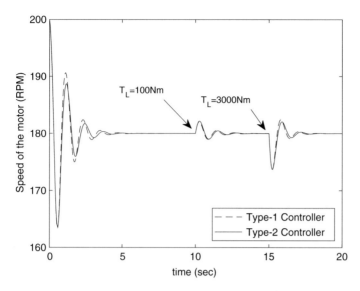

Fig. 5.7 The response of the closed-loop system controlling IM in the presence of variable load

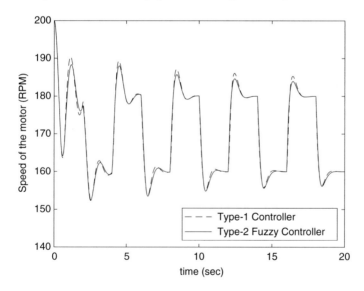

Fig. 5.8 The tracking response of the closed-loop control system

concluded from these simulations that when the system suffers from parameter variation, IT2FLS may be used while if the uncertainty which exists in the system is low, the type-1 fuzzy controller may be more appropriate because of its less computational requirements.

References

1. Bertsekas, D.P.: Nonlinear Programming. Athena Scientific, Belmont (1999)
2. Lin, P.-Z., Hsu, C.-F., Lee, T.-T.: Type-2 fuzzy logic controller design for buck dc-dc converters. In: The 14th IEEE International Conference on Fuzzy Systems, 2005. FUZZ'05, pp. 365–370. IEEE (2005)
3. Lin, P.-Z., Lin, C.-M., Hsu, C.-F., Lee, T.-T.: Type-2 fuzzy controller design using a sliding-mode approach for application to dc-dc converters. IEE Proceed.-Electric Power Appl. **152**(6), 1482–1488 (2005)
4. Sahin, M.E., Okumus, H.İ.: Fuzzy logic controlled buck-boost dc-dc converter for solar energy-battery system. In: 2011 International Symposium on Innovations in Intelligent Systems and Applications (INISTA). IEEE (2011), pp. 394–397
5. Guesmi, K., Hamzaoui, A., Zaytoon, J.: Fuzzy controller synthesis for a dc-dc converter. In: Proceedings of the 48th IEEE Conference on Decision and Control, 2009 Held Jointly with the 2009 28th Chinese Control Conference. CDC/CCC 2009, pp. 3106–3111. IEEE (2009)
6. Wai, R.-J., Shih, L.-C.: Adaptive fuzzy-neural-network design for voltage tracking control of a dc-dc boost converter. IEEE Trans. Power Electron. **27**(4), 2104–2115 (2012)
7. Kurokawa, F., Maruta, H., Ueno, K., Mizoguchi, T., Nakamura, A., Osuga, H.: A new digital control dc-dc converter with neural network predictor. In: Energy Conversion Congress and Exposition (ECCE), 2010 IEEE, pp. 522–526. IEEE (2010)
8. Ouali, M., Kamoun, M.B.: Field-oriented control induction machine and control by sliding mode. Simul. Pract. Theory **5**(2), 121–136 (1997)
9. Pucci, M.: Direct field oriented control of linear induction motors. Electric Power Syst. Res. **89**, 11–22 (2012)
10. Lin, F.J., Wang, D.H., Huang, P.K.: Fpga-based fuzzy sliding-mode control for a linear induction motor drive. IEE Proceed. - Electric Power Appl. **152**(5), 1137–1148 (2005). https://doi.org/10.1049/ip-epa:20050205
11. Barrero, F., Gonzalez, A., Torralba, A., Galvan, E., Franquelo, L.G.: Speed control of induction motors using a novel fuzzy sliding-mode structure. IEEE Trans. Fuzzy Syst. **10**(3), 375–383 (2002). https://doi.org/10.1109/TFUZZ.2002.1006440
12. Panchade, V., Chile, R., Patre, B.: A survey on sliding mode control strategies for induction motors. Ann. Rev. Control **37**(2), 289–307 (2013)
13. Kaur, R., Brar, G.: Induction motor drives-a literature review
14. Malla, S.G., Malla, J.M.R.: Direct torque control of induction motor with fuzzy controller: a review. Int. J. Emerg. Trends Electr. Electr. (IJETEE–ISSN: 2320-9569) **10**(3), 1–16 (2014)
15. Masumpoor, S., Yaghobi, H., Khanesar, M.A.: Adaptive sliding-mode type-2 neuro-fuzzy control of an induction motor. Expert Syst. Appl. **42**(19), 6635–6647 (2015)
16. Al-Sulaiti, M.K., Abu-Rub, H., Nounou, H., Al-Khuleifi, A., Al-Motawa, A., Iqbal, A.: Induction motor drive system with adaptive fuzzy logic controller. IFAC Proceed. Vol. **42**(13), 97–102 (2009), 14th IFAC Conference on Methods and Models in Automation and Robotics

Chapter 6
Adaptive Sliding-Mode Fuzzy Control Systems: Lyapunov Approach

This chapter deals with the adaptive design of fuzzy controllers based on the sliding-mode control law. As mentioned earlier, a challenge with regards to designing a sliding-mode controller is the necessity of having the nominal dynamics of the system. This requires a series of modeling prior to the control of the system. However, the modeling process of real-time systems requires complete knowledge of the dynamics of the system and its parameter values. Moreover, the modeling process includes simplifications and neglecting some phenomena such as nonlinearities, backlashes, friction, and time delays. Hence, the use of a fuzzy identifier in the structure of the controller makes it possible to control the system with less *a priori* knowledge about the dynamics of the system. In this chapter, the adaptive design of sliding-mode controllers using fuzzy identifiers is discussed in different cases. The stability analysis of the adaptation laws designed for the adaptive fuzzy systems is considered using appropriate Lyapunov function. The stability analysis is done in the presence of modeling error, unmodelled dynamics, and uncertainties.

6.1 Sliding-Mode Adaptive Type-1 Fuzzy Controller Design

In this section, sliding-mode adaptation laws are designed for T1FLSs. It is an indirect adaptive controller design when an identifier exists in the control structure whose output is used to design an appropriate control law. Figure 6.1 presents the block diagram of an indirect adaptive control system. In the design of sliding-mode controller, there are two distinct cases. The first case is when the coefficient of the control signal is a constant value and the second is when the coefficient of the control signal is itself a function of the states of the system. In this case, the former is considered

© Springer Nature Switzerland AG 2021   125
M. Ahmadieh Khanesar et al., *Sliding-Mode Fuzzy Controllers*, Studies in Systems,
Decision and Control 357, https://doi.org/10.1007/978-3-030-69182-0_6

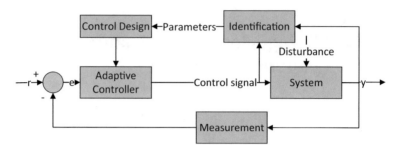

Fig. 6.1 Block diagram of indirect adaptive controllers

first as it is a rather simple case and then, the latter is paid much attention. Proportional integral action makes the steady state of the system more robust with respect to uncertainties and output disturbances. This type of fuzzy controller is considered in Sect. 6.1.3. Direct adaptive controllers do not necessitate an identifier and may require less computations and less memory. These kinds of controllers are investigated in Sect. 6.1.4. In the design of an adaptive fuzzy controller, it is recommended that the parameters of the antecedent part be optimized. Such optimization is more difficult as it may require Taylor expansion and suppression of higher-order terms in the Taylor expansion. In Sect. 6.1.5, the adaptation laws of the antecedent part along with those of consequent part are given. Terminal sliding-mode adaptive fuzzy controllers guarantee the finite-time reaching of the states of the system, which is highly desirable. This type of controller is investigated in Sect. 6.1.6.

6.1.1 Constant Control Signal Coefficient Case

Consider an nth-order nonlinear dynamic system as follows:

$$x^{(n)} = f(x, t) + u(t) + d(t), \tag{6.1}$$

where $x \in R^n$ is the state vector of the system, while $u \in R$ and $y \in R$ are the input and output of the system, respectively. The nonlinear time-varying function $f(x, t)$ is an unknown smooth function, which is to be identified using an appropriate T1FLS. The signal $d(t)$ represents unknown bounded external disturbance of the system, which satisfies $d(t) \leq D$. The upper bound D may or may not be known *a priori*. Too large an upper bound for the disturbance signal is undesirable because it may result in too large a control signal, which may cause chattering if it does not cause saturation in the actuators. In the current case, it is assumed that it is a known value. The reference signal is considered to be $r(t)$, and it is sufficiently smooth, i.e., its n consecutive derivatives do exist. Hence, the control objective is to make x, the states of the system, track this reference signal in the presence of model uncertainties and unknown disturbances.

Let the tracking error be:

$$e = x - r = (x - r, \dot{x} - \dot{r}, \ldots, x^{(n-1)} - r^{(n-1)}) = (e, \dot{e}, \ldots, e^{(n-1)}). \tag{6.2}$$

Then, a sliding surface is defined based on these error signals as follows:

$$s = e^{(n-1)} + \lambda_{n-1} e^{(n-2)} + \cdots + \lambda_2 \dot{e} + \lambda_1 e, \tag{6.3}$$

where the coefficients $\lambda_1, \ldots, \lambda_{n-1}$ are selected such that the following polynomial is a Hurwitz polynomial, i.e., real value of the roots of it is strictly negative $\xi^n + \lambda_{n-1} \xi^{n-1} + \cdots + \lambda_2 \xi^2 + \lambda_1 \xi$. In order to guarantee the finite-time convergence of the sliding manifold to *zero*, it is required to $s\dot{s} \leq -\eta |s|$.

It is highly desired that the states of the system be on the sliding surface, i.e., $s = 0$, and remain on this surface. In order to guarantee that the states of the system will remain on the sliding manifold, the time derivative of s must be equal to *zero*. This is called the sliding phase. The control signal which guarantees sliding phase, is called u_{eq} and is defined as follows. The time derivative of the s is obtained as follows:

$$\dot{s} = x^{(n)} - r^{(n)} + \lambda_{n-1} e^{(n-1)} + \cdots + \lambda_2 \ddot{e} + \lambda_1 \dot{e}. \tag{6.4}$$

By substituting $x^{(n)}$ from (6.1) to (6.4), the following equation is obtained.

$$\dot{s} = f(x, t) + u(t) + d(t) - r^{(n)} + \lambda_{n-1} e^{(n-1)} + \cdots + \lambda_2 \ddot{e} + \lambda_1 \dot{e}. \tag{6.5}$$

Hence, the control signal (u_{eq}) to maintain sliding phase is to be as follows:

$$u_{eq}(t) = -f(x, t) - d(t) + r^{(n)} - \sum_{i=1}^{n-1} \lambda_i e^{(i)}. \tag{6.6}$$

However, as mentioned earlier, $f(x, t)$, the dynamics of the system and its external disturbance ($d(t)$), are assumed to be unknown. Since $f(x, t)$ is a nonlinear smooth function, any of the general function approximators may be used to identify this function. T1FLS is known to be a general function approximator, so it has the ability to approximate any smooth function with a desirable approximation error, provided that sufficient number of rules are used in the construction of the fuzzy system. The selection of the parameters of the fuzzy system is important as it is assumed that little is known about these parameters. The stability and the performance of the control system depend on the selection of these parameters. Hence, the Lyapunov stability theory is used to adaptively tune these parameters based on the error and the sliding parameter and insure the stability of the closed loop system.

When the states of the system are not on the sliding manifold, i.e., $s \neq 0$, it is required that a switching control signal u_r be used to push the states of the system to the sliding manifold. Furthermore, this part of the control signal is responsible for the compensation of the error caused by time-varying external disturbances and modeling error in the system. The overall control signal u is defined as follows:

$$u(t) = u_{eq}(t) - u_r(t). \tag{6.7}$$

The T1FLS considered for the estimation of the nonlinear smooth function $f(x, t)$ is as follows:

$$\hat{f}(x|\theta_f) = \frac{\sum_{i=1}^{M} w_i \theta_{fi}}{\sum_{i=1}^{M} w_i}, \tag{6.8}$$

where w_i is the firing strength of the ith rule, θ_i is its corresponding consequent part parameters, and M is the number of rules. In order to simplify the following equations and avoid too many fractional equations, the parameter h_i is defined as the normalized value of the firing strength of the rules and is defined as follows:

$$h_i = \frac{w_i}{\sum_{i=1}^{M} w_i}. \tag{6.9}$$

In this case, T1FLS of (6.10) can be rewritten as follows:

$$\hat{f}(x|\theta_f) = \sum_{i=1}^{M} h_i \theta_{fi}. \tag{6.10}$$

By substituting the T1FLS of (6.10) in the sliding-mode control signal of (6.6), the overall sliding-mode control signal of (6.7) is obtained as follows:

$$u(t) = -\hat{f}(x|\theta_f) + r^{(n)} - \sum_{i=1}^{n-1} \lambda_i e^{(i)} - u_r(t). \tag{6.11}$$

Theorem 6.1 *Consider the nth-order nonlinear dynamic system of (6.1) with the control objective to track the reference signal r(t) with the sliding manifold as defined in (6.3). If the control signal (6.11) is applied with $\hat{f}(x|\theta_f)$, as given in (6.10), whose parameters are adaptively tuned as in (6.12) and u_r is defined as being equal to $-Ksgn(s)$, then sliding manifold converges to zero in finite time, resulting in sliding-mode behavior for the system. Hence, the tracking error of the system will asymptotically converge to zero, while the parameter errors remain bounded.*

$$\dot{\theta}_{fi} = \gamma_1 h_i s. \tag{6.12}$$

Proof The unknown optimal parameters of the T1FLS (θ_{fi}^*) are defined as follows:

$$\theta_{fi}^* = arg \min_{\theta_f \in S_f} \left[\sup_{x \in R^n} |f(x, t) - \hat{f}(x|\theta_f)| \right],$$ (6.13)

where S_f is a convex set to which the optimal values of T1FLS belong. Even if the optimal parameters of T1FLS were known, there exists a minimum functional approximation error between the T1FLS constructed by its optimal values and the real nonlinear function $f(x, t)$. This error is called the minimum functional approximation error (MFAE) [1] and is defined as follows:

$$\varepsilon_f = f(x, t) - \hat{f}(x|\theta_f^*).$$ (6.14)

From the general function approximation property of T1FLS, it is known that ε_f is bounded by a constant value as $|\varepsilon_f| < M_f$. Furthermore, it follows from the general function approximation property of T1FLS that the value of M_F for smooth functions can be made as small as desired by using sufficient number of rules for the fuzzy system.

We have the following equation for the time derivative of the sliding surface s.

$$\dot{s} = f(x, t) + u(t) + d(t) - r^{(n)} + \sum_{i=1}^{n-1} \lambda_i e^{(i)}.$$ (6.15)

Applying the control signal of (6.11)–(6.15), the following equation is obtained.

$$\dot{s} = f(x, t) + d(t) - \sum_{i=1}^{M} h_i \theta_i - u_r(t)$$

$$= f(x, t) + d(t) - \sum_{i=1}^{M} h_i \theta_i - u_r(t)$$

$$= f(x, t) + \hat{f}(x|\theta_f^*) - \hat{f}(x|\theta_f^*) + d(t) - \sum_{i=1}^{M} h_i \theta_i - u_r(t)$$

$$= \varepsilon_f + \hat{f}(x|\theta_f^*) + d(t) - \sum_{i=1}^{M} h_i \theta_i - u_r(t)$$

$$= \varepsilon_f + d(t) + \sum_{i=1}^{M} h_i \tilde{\theta}_i - u_r(t),$$ (6.16)

where $\tilde{\theta}_i$ is the error between the optimal parameters of the fuzzy system and its current estimated values as follows:

$$\tilde{\theta}_i = \theta_i^* - \theta_i. \tag{6.17}$$

In order to analyze the stability of the system, the following Lyapunov function is considered.

$$V = \frac{1}{2}s^2 + \frac{1}{2\gamma_1}\sum_{i=1}^{M}\tilde{\theta}_i^2, \tag{6.18}$$

where γ_1 is a positive constant. The time derivative of this Lyapunov function is obtained as follows:

$$\dot{V} = s\dot{s} + \frac{1}{\gamma_1}\sum_{i=1}^{M}\tilde{\theta}_i\dot{\theta}_i, \tag{6.19}$$

$$\dot{V} = s\left(\varepsilon_f + d(t) + \sum_{i=1}^{M}h_i\tilde{\theta}_i - u_r(t)\right) + \frac{1}{\gamma_1}\sum_{i=1}^{M}\tilde{\theta}_i\dot{\theta}_i$$

$$= s(\varepsilon_f + d(t) - u_r(t)) + \sum_{i=1}^{M}\tilde{\theta}_i\left(\frac{1}{\gamma_1}\dot{\theta}_i + h_i s\right). \tag{6.20}$$

Considering the adaptation law of (6.12), the following equation for the time derivative of the Lyapunov function is obtained.

$$\dot{V} = s(\varepsilon_f + d(t) - u_r(t))$$

$$\leq \xi_f|s(t)| + D|s(t)| - s(t)u_r(t). \tag{6.21}$$

Considering $u_r(t)$ as being equal to $K sign(s(t))$, the following equation is obtained.

$$\dot{V} \leq \xi_f|s(t)| + D|s(t)| - K|s(t)|. \tag{6.22}$$

It follows that

$$\dot{V} \leq -\eta|s(t)| \tag{6.23}$$

provided that $\eta + \xi \leq K$. Since the Lyapunov function V is positive definite and its time derivative $\dot{V}$ is negative semi-definite, the Lyapunov function V is decreasing and its upper bound is its initial value, i.e., $V \in L_\infty$. It follows that $s(t) \in L_\infty$ and $\tilde{\theta}_f \in L_\infty$. Since θ_fs are constant parameters, from $\tilde{\theta}_f \in L_\infty$, it is concluded that the parameters of the identifying fuzzy system remain bounded. Furthermore, from (6.23), it follows that

$$\int_0^\infty |s(\tau)|d\tau \leq \frac{V(0) - V(\infty)}{\eta} \tag{6.24}$$

which implies that $s(t) \in L_1$. Due to Barbalat's Lemma ($s \in L_p \cap L_\infty$ and $\dot{s} \in L_\infty$ $\Longrightarrow \lim_{t\to\infty} s(t) = 0$), it follows that the sliding parameter converges asymptotically to zero.

6.1.2 Nonlinear Control Signal Coefficient Case

It is also possible to design a similar adaptive sliding-mode fuzzy control system for when the dynamics of a nonlinear system includes an unknown nonlinear coefficient for the control signal. The system in this case is more general and is of the following form [2].

$$x^{(n)} = f(x, t) + g(x, t)u + d(t)$$
$$y = x, \tag{6.25}$$

where $f(x) : R^n \to R$ and $g(x) : R^n \to R$ are unknown smooth nonlinear functions of the states of the system, $d(t)$ is an unknown external disturbance of the system that satisfies $|d(t)| \leq D$, and y is the output of the system. For the nonlinear system of (6.25) to be controllable, it is required that $g(x) \neq 0$. Furthermore, it is required to assume that $g(x) > 0$. In addition, the states of the system must track the desired reference trajectory $r(t)$. Similar to the previous case, the sliding surface is defined as follows:

$$s(t) = e^{(n-1)}(t) + \lambda_{n-1}e^{(n-2)}(t) + \cdots + \lambda_2\dot{e}(t) + \lambda_1 e(t), \tag{6.26}$$

where $e(t)$ is the error signal, which is defined as $e(t) = x(t) - r(t)$. The sliding-mode control signal for the system is considered to be as follows:

$$u(t) = \frac{1}{\hat{g}(x|\theta_g)} \left[-\hat{f}(x|\theta_f) + \sum_{i=1}^{n-1} \lambda_i e^{(i)} - r^{(n)} - K\text{sign}(s) \right], \tag{6.27}$$

where $\hat{g}(x|\theta_g)$ and $\hat{f}(x|\theta_f)$ are T1FLSs that approximate $g(x)$ and $f(x)$, respectively. These fuzzy systems are defined as follows:

$$\hat{f}(x|\theta_f) = \sum_{i=1}^{M} h_{fi}\theta_{fi}, \tag{6.28}$$

$$\hat{g}(x|\theta_g) = \sum_{i=1}^{M} h_{gi}\theta_{gi}, \tag{6.29}$$

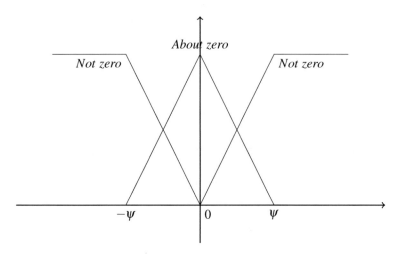

Fig. 6.2 The T1FLS considered for the fuzzy system designed in Sect. 6.1.2

where h_{fi} and h_{gi} are the normalized values of the firing strength of their corresponding fuzzy systems, with θ_{fi} and θ_{gi} being their consequent part parameters.

Furthermore, in order to avoid possible chattering in the system, the term to ensure the robustness of the system $-Ksign(s)$ is replaced by a single-input fuzzy system $l(s)$, which has two rules as follows:

$$IF \ s \ is \ About \ zero \ \ Then \ \ L = K_1 s$$
$$IF \ s \ is \ Not \ zero \ \ Then \ \ L = K_2 sign(s), \tag{6.30}$$

where K_1 and K_2 are two positive tunable parameters; the membership functions regarding *About zero* and *Not zero* are depicted in Fig. 6.2. The chattering in the control signal can be lessen by appropriate selection of the parameter ψ. While a small value for this parameter is desired to guarantee the stability and robustness of the system, too small a value may result in chattering in the control signal. The MF considered for the *About zero* fuzzy set is as follows:

$$\mu_z(s) = max\left(1 - \left|\frac{s}{\psi}\right|, 0\right) \tag{6.31}$$

while the MF for the *Not zero* fuzzy set is considered to be its complement and is defined as follows:

$$\mu_N(s) = 1 - \mu_z(s). \tag{6.32}$$

Hence, the output of T1FLS $l(s)$ is obtained as follows:

$$l(s) = K_1\mu_z(s) + K_2\mu_N(s)sign(s). \tag{6.33}$$

Using (6.33), the control signal is modified as follows:

$$u(t) = \frac{1}{\hat{g}(x|\theta_g)} \left[-\hat{f}(x|\theta_f) + \sum_{i=1}^{n-1} \lambda_i e^{(i)} - r^{(n)} + l(s) \right].$$

(6.34)

Theorem 6.2 *Consider the nth-order nonlinear dynamic system of (6.25) with reference signal r(t) with the sliding manifold as defined in (6.26). If the control signal (6.34) is applied to the system with $\hat{f}(x|\theta_f)$ and $g(x|\theta_g)$ are as given in (6.28) and (6.29) whose parameters are adaptively tuned as in (6.35) and (6.36), then sliding manifold converges to zero in finite time, resulting in sliding-mode behavior for the system. Hence, the tracking error of the system will asymptotically converge to zero while parameter errors remain bounded.*

$$\dot{\theta}_{fi} = \gamma_1 \, h_{fi} \, s,$$

(6.35)

$$\dot{\theta}_{gi} = \gamma_2 \, h_{gi} \, s \, u,$$

(6.36)

$$\dot{K}_1 = \begin{cases} \gamma_3 \, \mu_Z(s) \, s & If \; K_1 > 0 \\ & or \; K_1 = 0 \; and \; s > 0 \; , \\ 0 & Otherwise \end{cases}$$

(6.37)

$$\dot{K}_2 = \gamma_4 \, \mu_N(s) \, |s|.$$

(6.38)

Proof The unknown optimal parameters of the T1FLSs $\hat{f}(x|\theta_f)$ and $\hat{g}(x|\theta_g)$ namely θ_{fi}^* and θ_{gi}^*, are defined as follows:

$$\theta_{fi}^* = arg \min_{\theta_f \in S_f} \left[\sup_{x \in R^n} |f(x,t) - \hat{f}(x|\theta_f)| \right],$$

(6.39)

$$\theta_{gi}^* = arg \min_{\theta_g \in S_g} \left[\sup_{x \in R^n} |g(x,t) - \hat{g}(x|\theta_g)| \right],$$

(6.40)

where S_f and S_g are convex sets that include θ_{fi}^* and θ_{gi}^*. The minimum functional approximation error (MFAE) between $f(x)$ and $g(x)$ and their best approximations $\hat{f}(x|\theta_f)$ and $\hat{g}(x|\theta_g)$ are defined as follows:

$$\varepsilon_f = f(x) - \hat{f}(x|\theta_f^*),$$

(6.41)

$$\varepsilon_g = g(x) - \hat{g}(x|\theta_g^*)$$

(6.42)

which satisfy the following equations.

$$|\varepsilon_f| < M_f$$
$$|\varepsilon_g| < M_g,$$

where M_f and M_g are unknown positive constants, which can be made as small as desired using a sufficiently large number of rules for the fuzzy approximators. The time derivative of the sliding surface (6.26) is obtained as follows:

$$\dot{s} = f(x) + g(x)u(t) + d(t) - r^{(n)} + \sum_{i=1}^{n-1} \lambda_i e^{(i)}. \tag{6.43}$$

It follows from (6.34) that

$$\hat{g}(x|\theta_g)u(t) + \hat{f}(x|\theta_f) - \sum_{i=1}^{n-1} \lambda_i e^{(i)} + r^{(n)} - l(s) = 0. \tag{6.44}$$

By adding (6.44)–(6.43), the following equation is obtained.

$$\dot{s} = f(x) + g(x)u + d(t) - \sum_{i=1}^{M} h_{fi}\theta_{fi} - \sum_{i=1}^{M} h_{gi}\theta_{gi} - l(s)$$

$$= f(x) - \hat{f}(x|\theta_f^*) - \hat{g}(x|\theta_g^*)u + g(x)u + d(t) + \hat{f}(x|\theta_f^*) - \sum_{i=1}^{M} h_{fi}\theta_{fi}$$

$$+ \hat{g}(x|\theta_g^*)u - \sum_{i=1}^{M} h_{gi}\theta_{gi}u - l(s)$$

$$= f(x) - \sum_{i=1}^{M} h_{fi}\theta_{fi}^* + g(x)u - \sum_{i=1}^{M} h_{gi}\theta_{gi}^*u + d(t) + \sum_{i=1}^{M} h_{fi}\theta_{fi}^* - \sum_{i=1}^{M} h_{fi}\theta_{fi}$$

$$+ \sum_{i=1}^{M} h_{gi}\theta_{gi}^*u - \sum_{i=1}^{M} h_{gi}\theta_{gi}u - l(s)$$

$$= \varepsilon_f + \varepsilon_g u + d(t) + \sum_{i=1}^{M} h_{fi}\tilde{\theta}_{fi} + \sum_{i=1}^{M} h_{gi}\tilde{\theta}_{gi}u - l(s),$$

where $\tilde{\theta}_{fi}$ and $\tilde{\theta}_{gi}$ are defined as follows:

$$\tilde{\theta}_{fi} = \theta_{fi}^* - \theta_{fi}$$
$$\tilde{\theta}_{gi} = \theta_{gi}^* - \theta_{gi}. \tag{6.45}$$

In order to analyze the stability of the system, the following Lyapunov function is considered.

$$V = \frac{1}{2}s^2 + \frac{1}{2\gamma_1}\sum_{i=1}^{M}\tilde{\theta}_{fi}^2 + \frac{1}{2\gamma_2}\sum_{i=1}^{M}\tilde{\theta}_{gi}^2 + \frac{1}{2\gamma_3}\tilde{K}_1^2 + \frac{1}{2\gamma_4}\tilde{K}_2^2, \tag{6.46}$$

where γ_1, γ_2, γ_3, and γ_4 are positive constants acting as the learning rate of their corresponding parameters. Furthermore, the parameters $\tilde{K}_1$ and $\tilde{K}_2$ are defined so that they are equal to $\tilde{K}_1 = K_1 - K_1^*$ and $\tilde{K}_2 = K_2 - K_2^*$ with K_1^* and K_2^* being the optimal values of their corresponding parameters. The time derivative of this Lyapunov function (6.46) is obtained as follows:

$$\dot{V} = s\dot{s} + \frac{1}{\gamma_1}\sum_{i=1}^{M}\tilde{\theta}_{fi}\dot{\theta}_{fi} + \frac{1}{\gamma_2}\sum_{i=1}^{M}\tilde{\theta}_{gi}\dot{\theta}_{gi} + \frac{1}{\gamma_3}\tilde{K}_1\dot{K}_1 + \frac{1}{\gamma_4}\tilde{K}_2\dot{K}_2$$

$$= s\left(\varepsilon_f + \varepsilon_g u + d(t) + \sum_{i=1}^{M}h_{fi}\tilde{\theta}_{fi} + \sum_{i=1}^{M}h_{gi}\tilde{\theta}_{gi}u - l(s)\right)$$

$$+ \frac{1}{\gamma_1}\sum_{i=1}^{M}\tilde{\theta}_{fi}\dot{\theta}_{fi} + \frac{1}{\gamma_2}\sum_{i=1}^{M}\tilde{\theta}_{gi}\dot{\theta}_{gi} + \frac{1}{\gamma_3}\tilde{K}_1\dot{K}_1 + \frac{1}{\gamma_4}\tilde{K}_2\dot{K}_2$$

$$= \sum_{i=1}^{M}\tilde{\theta}_{fi}\left(h_{fi}s + \gamma_1^{-1}\dot{\theta}_{fi}\right) + \sum_{i=1}^{M}\tilde{\theta}_{gi}\left(h_{gi}su + \gamma_2^{-1}\dot{\theta}_{gi}\right) - l(s)s$$

$$+ \frac{1}{\gamma_3}\tilde{K}_1\dot{K}_1 + \frac{1}{\gamma_4}\tilde{K}_2\dot{K}_2 + s\left(\varepsilon_f + \varepsilon_g u + d(t)\right). \tag{6.47}$$

Considering the adaptation laws of (6.35) and (6.36), one obtains the following equation for the time derivative of the Lyapunov function.

$$\dot{V} = -l(s)s + \frac{1}{\gamma_3}\tilde{K}_1\dot{K}_1 + \frac{1}{\gamma_4}\tilde{K}_2\dot{K}_2 + s\left(\varepsilon_f + \varepsilon_g u + d(t)\right). \tag{6.48}$$

By substituting $l(s)$ as in (6.33)–(6.48), the following equation is obtained.

$$\dot{V} = -K_1\mu_Z(s)s - K_2\mu_N(s)|s| + \frac{1}{\gamma_3}\tilde{K}_1\dot{K}_1 + \frac{1}{\gamma_4}\tilde{K}_2\dot{K}_2 + s\left(\varepsilon_f + \varepsilon_g u + d(t)\right)$$

$$= -\eta|s| + \eta|s| - K_1\mu_Z(s)s - K_2\mu_N(s)|s| + \frac{1}{\gamma_3}\tilde{K}_1\dot{K}_1 + \frac{1}{\gamma_4}\tilde{K}_2\dot{K}_2$$

$$+ s\left(\varepsilon_f + \varepsilon_g u + d(t)\right). \tag{6.49}$$

Since ε_g has a small value, it is possible to consider $|\varepsilon_f + \varepsilon_g u| \le v$. Hence, one obtains

$$\dot{V} \le -\eta|s| + \eta|s| - K_1\mu_Z(s)s - K_2\mu_N(s)|s| + \frac{1}{\gamma_3}\tilde{K}_1\dot{K}_1 + \frac{1}{\gamma_4}\tilde{K}_2\dot{K}_2 + (v + D)|s|$$

and considering $v + D \le \eta$, it follows that

$$\dot{V} \leq \eta|s| - K_1\mu_Z(s)s - K_2\mu_N(s)|s| + \frac{1}{\gamma_3}\tilde{K}_1\dot{K}_1 + \frac{1}{\gamma_4}\tilde{K}_2\dot{K}_2$$

$$= \eta|s| - K_1^*\mu_Z(s)s - K_2^*\mu_N(s)|s| + \tilde{K}_1\left(\frac{1}{\gamma_3}\dot{K}_1 - \mu_Z(s)s\right) + \left(\frac{1}{\gamma_4}\dot{K}_2 - \mu_N(s)|s|\right).$$

When $K_1 > 0$ or $K_1 = 0$ and $\mu_Z(s)s > 0$, the adaptation laws of (6.37) and (6.38) results in the following inequality for the time derivative of the Lyapunov function.

$$\dot{V} \leq \eta|s| - K_1^*\mu_Z(s)s - K_2^*\mu_N(s)|s|. \tag{6.50}$$

On the other hand, when $K_1 = 0$ and $\mu_Z(s)s > 0$, it follows from (6.37) that $\dot{K}_1 = 0$ and considering the fact that K_1^* has a positive value, one obtains $-\tilde{K}_1\mu_z(s)s \leq 0$. Hence, in both cases, we have the following inequality:

$$\dot{V} \leq \eta|s| - K_1^*\mu_Z(s)s - K_2^*\mu_N(s)|s|. \tag{6.51}$$

When $|s| > \psi$, we have

$$\dot{V} \leq \eta|s| - K_2^*|s|. \tag{6.52}$$

It is sufficient to consider K_2^* as follows:

$$2\eta < K_2^* \tag{6.53}$$

to obtain $\dot{V} \leq -\eta|s|$, which guarantees asymptotic stability of the sliding surface. On the other hand, when $|s| < \psi$, the following inequality is obtained:

$$\dot{V} \leq \eta|s| - K_1^*\mu_Z(s)s - K_2^*\mu_N(s)|s|. \tag{6.54}$$

In this case, the sign of time derivative of the Lyapunov function is undefined. Hence, the sliding surface converges to a small neighborhood of *zero*, in which $|s| \leq \psi$ and it remains there.

6.1.3 Adding PI to Sliding-Mode Fuzzy Controller

PI-type controllers are widely known to be able to lessen and even completely eliminate the steady-state error. It has previously used in parallel to adaptive fuzzy controllers to provide a mechanism to deal with large and fast disturbances without the usage of *sign* function, which may result in chattering [3].

The system to be controlled is considered to be of the following form.

$$x^{(n)}(t) = f(x(t)) + g(x(t))u + d(t)$$
$$y(t) = x(t),$$ (6.55)

where the definitions of the nonlinear functions $f(x)$, $g(x)$ and the disturbance signal $d(t)$ are the same as the definitions of $d(t)$ as a bounded signal and $|d(t)| < D$ and the sliding manifold is the same as (6.26). The PI-type sliding-mode control signal for the system is considered to be as follows:

$$u(t) = \frac{1}{\hat{g}(x|\theta_g)}\left[-\hat{f}(x|\theta_f) + \sum_{i=1}^{n-1} \lambda_i e^{(i)} - r^{(n)} - K_P sat(s/\phi) - K_I \int_0^t s(\tau)d\tau \right],$$ (6.56)

where $\hat{g}(x|\theta_g)$ and $\hat{f}(x|\theta_f)$ are as defined in (6.28) and (6.29), respectively. The saturation function $sat(.)$ is defined as follows:

$$sat(s/\Phi) = \begin{cases} s/\Phi & if \ |s| < \Phi \\ sign(s) & Otherwise \end{cases}.$$ (6.57)

Theorem 6.3 *Consider the nth-order nonlinear dynamic system of (6.55) with reference signal $r(t)$ with the sliding manifold as defined in (6.26). The control signal (6.56) is applied to the system which includes two fuzzy approximators of $\hat{f}(x|\theta_f)$ and $g(x|\theta_g)$, which are given in (6.28) and (6.29) and u_{PI} is the PI part of the control signal. The parameters of the fuzzy systems are adaptively tuned as in (6.58) and (6.59) and the parameters of the PI controller are adjusted as in (6.60) and (6.61). Then the sliding manifold converges to a neighborhood near zero.*

$$\dot{\theta}_{fi} = \gamma_1 h_{fi} s,$$ (6.58)

$$\dot{\theta}_{gi} = \gamma_2 h_{gi} s u,$$ (6.59)

$$\dot{K}_P = \gamma_3 s(t) sat(s(t)/\Phi),$$ (6.60)

$$\dot{K}_I = \gamma_4 s \int_0^t s(\tau)d\tau.$$ (6.61)

Proof The unknown optimal parameters θ_{fi}^* and θ_{gi}^* are as defined in (6.39) and (6.40) and the MAFE signals ε_f and ε_g are as defined in (6.41)–(6.42). The time derivative of the sliding surface (6.26) is obtained as follows:

$$\dot{s} = f(x) + g(x)u(t) + d(t) - r^{(n)} + \sum_{i=1}^{n-1} \lambda_i e^{(i)}.$$ (6.62)

It follows from (6.56) that

$$\hat{g}(x|\theta_g)u(t) + \hat{f}(x|\theta_f) + \sum_{i=1}^{n-1} \lambda_i e^{(i)} - r^{(n)} + K_P sat(s/\phi) + K_I \int_0^t s(\tau)d\tau = 0.$$

$$(6.63)$$

By adding (6.63)–(6.62), the following equation is obtained:

$$\dot{s} = f(x) + g(x)u + d(t) - \hat{g}(x|\theta_g)u(t) - \hat{f}(x|\theta_f) - K_P sat(s/\phi) - K_I \int_0^t s(\tau)d\tau$$

$$= f(x) - \sum_{i=1}^{M} h_{fi}\theta_{fi}^* + g(x)u - \sum_{i=1}^{M} h_{gi}\theta_{gi}^* u + d(t) + \sum_{i=1}^{M} h_{fi}\tilde{\theta}_{fi} + \sum_{i=1}^{M} h_{gi}\tilde{\theta}_{gi}u$$

$$- K_P sat(s/\phi) - K_I \int_0^t s(\tau)d\tau$$

$$= \varepsilon_f + \varepsilon_g u + d(t) + \sum_{i=1}^{M} h_{fi}\tilde{\theta}_{fi} + \sum_{i=1}^{M} h_{gi}\tilde{\theta}_{gi}u - K_P sat(s/\phi) - K_I \int_0^t s(\tau)d\tau,$$

where $\tilde{\theta}_{fi}$ and $\tilde{\theta}_{gi}$ are defined as follows:

$$\tilde{\theta}_{fi} = \theta_{fi}^* - \theta_{fi}$$

$$\tilde{\theta}_{gi} = \theta_{gi}^* - \theta_{gi}.$$

$$(6.64)$$

In order to analyze the stability of the system, the following Lyapunov function is considered.

$$V = \frac{1}{2}s^2 + \frac{1}{2\gamma_1}\sum_{i=1}^{M}\tilde{\theta}_{fi}^2 + \frac{1}{2\gamma_2}\sum_{i=1}^{M}\tilde{\theta}_{gi}^2 + \frac{1}{2\gamma_3}\tilde{K}_P^2 + \frac{1}{2\gamma_4}K_I^2,$$

$$(6.65)$$

where $\gamma_1, \gamma_2, \gamma_3$, and γ_4 are positive constants. Moreover, the parameter $\tilde{K}_P$ is defined as being equal to $\tilde{K}_P = K_P - K_P^*$, with K_P^* being the optimal value of the proportional part of the PI controller. The time derivative of this Lyapunov function (6.65) is obtained as follows:

$$\dot{V} = s\dot{s} + \frac{1}{\gamma_1}\sum_{i=1}^{M}\tilde{\theta}_{fi}\dot{\theta}_{fi} + \frac{1}{\gamma_2}\sum_{i=1}^{M}\tilde{\theta}_{gi}\dot{\theta}_{gi} + \frac{1}{\gamma_3}\tilde{K}_P\dot{K}_P + \frac{1}{\gamma_4}\tilde{K}_I\dot{K}_I$$

$$= s\left(\varepsilon_f + \varepsilon_g u + d(t) + \sum_{i=1}^{M}h_{fi}\tilde{\theta}_{fi} + \sum_{i=1}^{M}h_{gi}\tilde{\theta}_{gi}u - K_P sat(s/\phi) - K_I \int_0^t s(\tau)d\tau\right)$$

$$+ \frac{1}{\gamma_1}\sum_{i=1}^{M}\tilde{\theta}_{fi}\dot{\theta}_{fi} + \frac{1}{\gamma_2}\sum_{i=1}^{M}\tilde{\theta}_{gi}\dot{\theta}_{gi} + \frac{1}{\gamma_3}\tilde{K}_P\dot{K}_P + \frac{1}{\gamma_4}K_I\dot{K}_I$$

$$= \sum_{i=1}^{M}\tilde{\theta}_{fi}\left(h_{fi}s + \gamma_1^{-1}\dot{\theta}_{fi}\right) + \sum_{i=1}^{M}\tilde{\theta}_{gi}\left(h_{gi}su + \gamma_2^{-1}\dot{\theta}_{gi}\right) - s\left(K_P sat(s/\Phi) + K_I \int_0^t s(\tau)d\tau\right)$$

$$+ \frac{1}{\gamma_3}\tilde{K}_P\dot{K}_P + \frac{1}{\gamma_4}K_I\dot{K}_I + s\left(\varepsilon_f + \varepsilon_g u + d(t)\right). \tag{6.66}$$

Considering the adaptation laws of (6.58), (6.59), and (6.61), the following equations are obtained.

$$\dot{V} = -s(K_P sat(s/\Phi) + \frac{1}{\gamma_3}\tilde{K}_P\dot{K}_P + s\left(\varepsilon_f + \varepsilon_g u + d(t)\right)$$

$$= -K_P^*|s| - K_P^* s sat(s/\Phi) + K_P^*|s| - \tilde{K}_P s sat(s/\Phi) + \frac{1}{\gamma_3}\tilde{K}_P\dot{K}_P$$

$$+ s\left(\varepsilon_f + \varepsilon_g u + d(t)\right)$$

$$= -K_P^*|s| + W(s) - \tilde{K}_P s sat(s/\Phi) + \frac{1}{\gamma_3}\tilde{K}_P\dot{K}_P$$

$$+ s\left(\varepsilon_f + \varepsilon_g u + d(t)\right), \tag{6.67}$$

where

$$W(s) = K_P^*|s| - K_P^* s sat(s/\Phi). \tag{6.68}$$

It is to be noted that the function $W(s)$ is equal to *zero* when $|s| > \Phi$. Considering the adaptation law of K_P as in (6.60), the Eq. (6.67) can be further modified as follows:

$$\dot{V} = -K_P^*|s| + W(s) + s\left(\varepsilon_f + \varepsilon_g u + d(t)\right). \tag{6.69}$$

If K_P^* is taken as $K_P^* > \eta + D + M_\varepsilon$ where $|\varepsilon_f + \varepsilon_g u| < M_\varepsilon$, we have the following equation:

$$\dot{V} = -\eta|s| + W(s) \tag{6.70}$$

which ensures that the sliding manifold converges to a neighborhood near *zero*, in which, $|s| \leq \Phi$ and remains there.

Remark It is to be noted that Φ is a design parameter, which can be considered as small as desired to make the upper bound of the sliding manifold as small as desired.

6.1.4   Sliding-Mode Direct Adaptive Fuzzy Control

It is a direct adaptive controller when no explicit plant model identification is required in order to generate the feedback control signal and the parameters of the controller are directly tuned from the input/output and the error signals collected from the system [4].

The first main difference between direct adaptive and indirect adaptive controllers is that no explicit mathematical model of the plant is needed when designing a direct adaptive controller. The second difference is that the identification error is used for the tuning of the parameters of an indirect adaptive controller, while tracking error is used to adjust the parameters of a direct adaptive controller. The block diagram of a direct adaptive controller is depicted in Fig. 6.3.

It is possible to design a direct sliding-mode fuzzy controller. The nonlinear dynamic system which is considered in this case is different from (6.25) in the constraints required by the Lyapunov function is considered in this case.

$$x^{(n)} = f(x) + g(x)u$$
$$y = x, \tag{6.71}$$

where $f(x) : R^n \to R$ and $g(x) : R^n \to R$ are sufficiently smooth nonlinear functions. It is further assumed that $0 < g_l < g(x) < g_u$ and $|\dot{g}(x)| < M_{\dot{g}}$. Let the tracking error be

$$e = x - r = (x - r, \dot{x} - \dot{r}, \dots, x^{(n-1)} - r^{(n-1)}) = (e, \dot{e}, \dots, e^{(n-1)}). \tag{6.72}$$

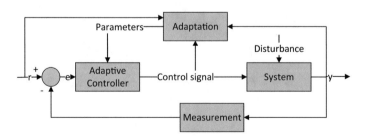

Fig. 6.3 Block diagram of direct adaptive controllers

Then, a sliding surface is defined based on these error signals as follows:

$$s = e^{(n-1)} + \lambda_{n-1}e^{(n-2)} + \cdots + \lambda_2 \dot{e} + \lambda_1 e, \tag{6.73}$$

where the coefficients $\lambda_1, \ldots, \lambda_{n-1}$ are coefficients of the Hurwitz polynomial. Considering (6.73), we have the following equation:

$$\dot{s} = f(x) + g(x)u + d(t) - r^{(n)} + \lambda_{n-1}e^{(n-1)} + \cdots + \lambda_2 \ddot{e} + \lambda_1 \dot{e}. \tag{6.74}$$

The sliding-mode control signal to guarantee the stability of the system in this case is as follows:

$$u^* = \frac{1}{g(x)}\left(-f(x) - \rho(t)\right), \tag{6.75}$$

where $\rho(t)$ is defined as follows:

$$\rho(t) = -r^{(n)} + \lambda_{n-1}e^{(n-1)} + \cdots + \lambda_2 \ddot{e} + \lambda_1 \dot{e} + K_1 sign(s) + K_2 s. \tag{6.76}$$

It is assumed that this control signal is constructed based on a fuzzy system whose IF–THEN rules are as follows:

$$R^i : \text{ If } x_1 \text{ is } A_{1i}, \ldots, x_n A_{ni} \text{ Then } u(t) = \theta_{1i} + \theta_{2i}\rho(t). \tag{6.77}$$

The output of such fuzzy system is inferred as follows:

$$u(t) = \hat{m}(x|\theta_{1i}, \theta_{2i}) = \sum_{i=1}^{M}\left(h_i\left(\theta_{1i} + \theta_{2i}\rho\right)\right). \tag{6.78}$$

It is assumed that the sliding-mode control signal of (6.75) can be approximated by this fuzzy system as follows:

$$u^*(t) = \sum_{i=1}^{M}\left(h_i\left(\theta_{1i}^* + \theta_{2i}^*\rho(t)\right)\right) + \varepsilon, \tag{6.79}$$

where the parameters θ_{1i}^* and θ_{2i}^* are the optimal parameters of the fuzzy system, which are defined as follows:

$$\theta_{1i}^*, \theta_{2i}^* = arg \min_{\theta_{1i}^*, \theta_{2i}^* \in S_m}\left[\sup_{x \in R^n}|\hat{m}(x|\theta_{1i}^*, \theta_{2i}^*) - \hat{m}(x|\theta_{1i}, \theta_{2i})|\right]. \tag{6.80}$$

Substituting $\rho(t)$ as in (6.76)–(6.74) yields the following equation:

$$\dot{s} = f(x) + g(x)u + d(t) + \rho(t) - K_1 sign(s) - K_2 s \tag{6.81}$$

and furthermore,

$$\dot{s} = f(x) + g(x)u - g(x)u^* + g(x)u^* + d(t) - K_1 sign(s) - K_2 s + \rho(t). \quad (6.82)$$

From (6.75), we have

$$f(x) + g(x)u^* + \rho(t) = 0. \quad (6.83)$$

Considering (6.83), Eq. (6.82) can be modified as follows:

$$\dot{s} = g(x)u - g(x)u^* + d(t) - K_1 sign(s) - K_2 s + \varepsilon. \quad (6.84)$$

By substituting u^* and u as in (6.79) and (6.78), Eq. (6.84) can be modified as follows:

$$\dot{s} = g(x)\left(\sum_{i=1}^{M}\left(\tilde{\theta}_{1i}h_i + \tilde{\theta}_{2i}h_i\rho(t)\right)\right) + d(t) - K_1 sign(s) - K_2 s + \varepsilon. \quad (6.85)$$

In order to analyze the stability of the controller, the following Lyapunov function is considered. This Lyapunov function guarantees the asymptotic stability of sliding surface s as well as the boundedness of the parameters of the direct sliding-mode fuzzy controller.

$$V = \frac{1}{2g(x)}s^2 + \frac{1}{2\gamma_1}\sum_{i=1}^{M}\tilde{\theta}_{1i}^2 + \frac{1}{2\gamma_2}\sum_{i=1}^{M}\tilde{\theta}_{2i}^2. \quad (6.86)$$

The time derivative of the Lyapunov function (6.86) is obtained as follows:

$$\dot{V} = \frac{1}{g(x)}s\dot{s} - \frac{\dot{g}(x)}{g^2(x)}s^2 + \frac{1}{\gamma_1}\sum_{i=1}^{M}\tilde{\theta}_{1i}\dot{\theta}_{1i} + \frac{1}{\gamma_2}\sum_{i=1}^{M}\tilde{\theta}_{2i}\dot{\theta}_{2i}. \quad (6.87)$$

By substituting the time derivative of the sliding manifold as in (6.85)–(6.87), the following equation for the time derivative of the Lyapunov function is obtained.

$$\dot{V} = \frac{1}{g(x)}s\left(g(x)\left(\sum_{i=1}^{M}\left(\tilde{\theta}_{1i}h_i + \tilde{\theta}_{2i}h_i\rho(t)\right)\right) + d(t) - K_1 sign(s) - K_2 s + \varepsilon\right)$$

$$- \frac{\dot{g}(x)}{g^2(x)}s^2 + \frac{1}{\gamma_1}\sum_{i=1}^{M}\tilde{\theta}_{1i}\dot{\theta}_{1i} + \frac{1}{\gamma_2}\sum_{i=1}^{M}\tilde{\theta}_{2i}\dot{\theta}_{2i}. \quad (6.88)$$

Considering the adaptation laws of the parameters of the fuzzy system θ_{1i} and θ_{2i} as

$$\dot{\theta}_{1i} = \gamma_1 sh_i$$
$$\dot{\theta}_{2i} = \gamma_2 sh_i\rho \quad (6.89)$$

Equation (6.88) can be rewritten as follows:

$$\dot{V} = \frac{1}{g(x)} s \left(d(t) - K_1 sign(s) - K_2 s + \varepsilon \right) - \frac{\dot{g}(x)}{g^2(x)} s^2. \tag{6.90}$$

Considering $K_1 > D + \eta g_u + M_\varepsilon$ $K_2 > \frac{M_{\dot{g}}}{g_l}$, the following equation is obtained.

$$\dot{V} \leq -\eta |s| \tag{6.91}$$

which guarantees the asymptotic stability of the sliding manifold using Barbalat's lemma. Furthermore, the boundedness of the parameters of the fuzzy system is also ensured.

6.1.5 Tuning Antecedent Part Parameters

In order to obtain superior performance, in this section, the tuning of the parameters of the antecedent part of fuzzy system is considered. The parameters of the antecedent appear nonlinearly in the output of the fuzzy system; so, the training of these parameters is not straight forward. In this section, using Taylor series expansion, the aforementioned adaptation laws are extended to the antecedent part parameters. In this case, it is assumed that the more general nonlinear dynamic system includes an unknown nonlinear coefficient for the control signal. The dynamic equation of the system, which is repeated here for ease of access, is as follows:

$$x^{(n)} = f(x) + b(x)w + d(t)$$
$$y = x, \tag{6.92}$$

where $f(x) : R^n \to R$ and $b(x) : R^n \to R$ are unknown smooth nonlinear functions of the system states and we have $0 < b(x) < b_{max}$, $d(t)$ is unknown external disturbance of the system, which satisfies $|d(t)| \leq D$ and w and y are the system input and output, respectively. It is possible to change the input of the system as

$$w = \frac{u}{2b_{max}} \tag{6.93}$$

in (6.92), which results in the following equation:

$$x^{(n)} = f(x) + g(x)u + d(t)$$
$$y = x. \tag{6.94}$$

In this case, we have $0 < g(x) \leq 0.5$. The sliding manifold is defined as follows:

$$s(t) = e^{(n-1)}(t) + \lambda_{n-1} e^{(n-2)}(t) + \cdots + \lambda_2 \dot{e}(t) + \lambda_1 e(t) \tag{6.95}$$

with $e(t)$ being the tracking error, which is defined as $e(t) = x(t) - r(t)$. Similar to the previous indirect controller design case, two T1FLSs are used to approximate the nonlinear functions, which exist in the dynamic of the system, namely $f(x)$ and $g(x)$. However, unlike similar cases in which only the consequent part parameters were tunable, the antecedent part parameters are considered to be adaptive and their corresponding adaptation laws are extracted from an appropriate Lyapunov function. The two T1FLSs considered in this case are as follows:

$$\hat{f}(x|\theta_f, \alpha_f) = \sum_{i=1}^{M} h_{fi}(x|\alpha_{fi})\theta_{fi}, \tag{6.96}$$

$$\hat{g}(x|\theta_g, \alpha_g) = \sum_{i=1}^{M} h_{gi}(x|\alpha_{gi})\theta_{gi}, \tag{6.97}$$

where h_{fi} and h_{gi} are the normalized values of the firing strength of their corresponding fuzzy rules with θ_{fi} and θ_{gi} being their tunable consequent part parameters. Furthermore, the parameters α_{fi} and α_{gi} are vectors of the antecedent part parameters with appropriate dimensions, which are among the tunable parameters in this case.

The sliding-mode control signal for the system is considered to be as follows:

$$u(t) = \frac{1}{\hat{g}(x|\theta_g, \alpha_g)}\left[-\hat{f}(x|\theta_f, \alpha_f) + \sum_{i=1}^{n-1} \lambda_i e^{(i)} - r^{(n)} - K\text{sign}(s)\right]. \tag{6.98}$$

Theorem 6.4 *Consider the nth-order nonlinear dynamic system of (6.92) with reference signal $r(t)$ with the sliding manifold as defined in (6.95). If the control signal (6.98) applied to the system with $\hat{f}(x|\theta_f)$ and $g(x|\theta_g)$ are as given in (6.96) and (6.97) whose parameters are adaptively tuned as in (6.99)–(6.103), then the sliding manifold converges to zero in finite time, resulting in sliding-mode behavior for the system. Hence, the tracking error of the system will asymptotically converge to zero while the parameter errors remain bounded.*

$$\dot{\theta}_{fi} = \gamma_1 h_{fi} s, \tag{6.99}$$

$$\dot{\theta}_{gi} = \gamma_2 h_{gi} s u, \tag{6.100}$$

$$\dot{K}_1 = \begin{cases} \gamma_3 \mu_Z(s) s & \text{If } K_1 > 0 \\ & \text{or } K_1 = 0 \text{ and } s > 0 \ , \\ 0 & \text{Otherwise} \end{cases} \tag{6.101}$$

$$\dot{K}_2 = \gamma_4 \mu_N(s) |s|, \tag{6.102}$$

$$\dot{\alpha}_{fi} = -\dot{\tilde{\alpha}}_{fi} = \gamma_2 \theta_{fi} \frac{\partial h_{fi}(x)}{\partial \alpha_{fi}(x)} s$$

$$\dot{\alpha}_{gi} = -\dot{\tilde{\alpha}}_{gi} = \gamma_4 \theta_{gi} \frac{\partial h_{gi}(x)}{\partial \alpha_{gi}(x)} su. \tag{6.103}$$

Proof It is assumed that the unknown optimal parameters of the T1FLSs $\hat{f}(x|\theta_f, \alpha_f)$ and $\hat{g}(x|\theta_g, \alpha_g)$, namely, θ_{fi}^*, α_{fi}^*, θ_{gi}^*, and θ_{gi}^*, are defined as follows:

$$\theta_{fi}^*, \alpha_{fi}^* = arg \min_{\theta_f \in S_f} \left[\sup_{x \in R^n} |f(x, t) - \hat{f}(x|\theta_f, \alpha_f)| \right], \tag{6.104}$$

$$\theta_{gi}^*, \alpha_{gi}^* = arg \min_{\theta_g \in S_g} \left[\sup_{x \in R^n} |g(x, t) - \hat{g}(x|\theta_g, \alpha_g)| \right], \tag{6.105}$$

where S_f and S_g are convex sets that contain the optimal parameters θ_{fi}^*, α_{fi}^*, θ_{gi}^*, and θ_{gi}^*. The MFAE between $f(x)$ and $g(x)$ and their best approximations $\hat{f}(x|\theta_f, \alpha_f)$ and $\hat{g}(x|\theta_g, \alpha_g)$ are defined as follows:

$$\varepsilon_f = f(x) - \hat{f}(x|\theta_f^*, \alpha_f^*), \tag{6.106}$$

$$\varepsilon_g = g(x) - \hat{g}(x|\theta_g^*, \alpha_g^*). \tag{6.107}$$

Considering the general function approximation property of T1FLSs, the following assumptions can be made:

$$|\varepsilon_f| < M_f$$
$$|\varepsilon_g| < M_g$$

with M_f and M_g being unknown positive constants which can be made as small as desired using sufficiently large number of rules for the fuzzy approximators. The time derivative of the sliding surface (6.95) is obtained as follows:

$$\dot{s} = f(x) + g(x)u(t) + d(t) - r^{(n)} + \sum_{i=1}^{n-1} \lambda_i e^{(i)}. \tag{6.108}$$

It follows from (6.98) that

$$\hat{g}(x|\theta_g, \alpha_g)u(t) + \hat{f}(x|\theta_f, \alpha_f) - \sum_{i=1}^{n-1} \lambda_i e^{(i)} + r^{(n)} + K \, sign(s) = 0. \tag{6.109}$$

By adding (6.109)–(6.108), the following equation is obtained:

$$\dot{s} = f(x) + g(x)u + d(t) - \sum_{i=1}^{M} h_{fi}(x|\alpha_{fi})\theta_{fi} - \sum_{i=1}^{M} h_{gi}(x|\alpha_{gi})\theta_{gi} - K\,sign(s)$$

$$= f(x) - \hat{f}(x|\theta_f^*, \alpha_f^*) + g(x)u - \hat{g}(x|\theta_g^*, \alpha_g^*)u + d(t) + \hat{f}(x|\theta_f^*, \alpha_f^*) - \sum_{i=1}^{M} h_{fi}(x|\alpha_{fi})\theta_{fi}$$

$$+ \hat{g}(x|\theta_g^*, \alpha_g^*)u - \sum_{i=1}^{M} h_{gi}(x|\alpha_{gi})\theta_{gi}u - K\,sign(s)$$

$$= f(x) - \hat{f}(x|\theta_f^*, \alpha_f^*) + g(x)u - \hat{g}(x|\theta_g^*, \alpha_g^*)u + d(t) + \sum_{i=1}^{M} h_{fi}(x|\alpha_{fi}^*)\theta_{fi}^*$$

$$- \sum_{i=1}^{M} h_{fi}(x|\alpha_{fi})\theta_{fi} + \hat{g}(x|\theta_g^*, \alpha_g^*)u - \sum_{i=1}^{M} h_{gi}(x|\alpha_{gi})\theta_{gi}u - K\,sign(s).$$

The firing strength of the fuzzy rules $h(x|\alpha_{gi}^*)$ around α_{gi} can be expanded as follows:

$$h_{gi}(x|\alpha_{gi}^*) = h_{gi}(x|\alpha_{gi}) + (\alpha_{gi}^* - \alpha_{gi})^T \left. \frac{\partial h_{gi}(x)}{\partial \alpha_{gi}} \right|_{\alpha_{gi}=\alpha_{gi}} + R_g, \qquad (6.110)$$

where R_g represents the higher-order terms of Taylor expansion of $h(x|\alpha_{gi}^*)$. Considering Gaussian MFs for the fuzzy system, it follows that $|R_g| < M_g$ with M_g having a positive value. Similarly, the firing strength of the fuzzy rules $h_{fi}(x|\alpha_{fi}^*)$ around α_{fi}^* can be expanded as follows:

$$h_{fi}(x|\alpha_{fi}^*) = h_{fi}(x|\alpha_{fi}) + (\alpha_{fi}^* - \alpha_{fi})^T \left. \frac{\partial h_{fi}(x)}{\partial \alpha_{fi}} \right|_{\alpha_{fi}=\alpha_{fi}} + R_f, \qquad (6.111)$$

where R_f represents the higher-order terms of Taylor expansion of $h(x|\alpha_{fi})$. Considering Gaussian MFs for the fuzzy system, it follows that $|R_f| < M_f$ with M_f having a positive value. Considering (6.110) and (6.111), the Eq. (6.110) can be further modified as follows:

$$\dot{s} = \varepsilon_f + \varepsilon_g u + d(t) + \sum_{i=1}^{M} h_{fi}(x|\alpha_{fi}^*)\theta_{fi}^*$$

$$- \sum_{i=1}^{M} h_{fi}(x|\alpha_{fi})\theta_{fi} + \sum_{i=1}^{M} h_{gi}(x|\alpha_{gi}^*)\theta_{gi}^*u - \sum_{i=1}^{M} h_{gi}(x|\alpha_{gi})\theta_{gi}u - K\,sign(s),$$

$$\dot{s} = \varepsilon_f + \varepsilon_g u + d(t) + \sum_{i=1}^{M} h_{fi}(x|\alpha_{fi}^*)\theta_{fi}^* + \sum_{i=1}^{M} h_{fi}(x|\alpha_{fi})\theta_{fi}^* - \sum_{i=1}^{M} h_{fi}(x|\alpha_{fi})\theta_{fi}^*$$

$$- \sum_{i=1}^{M} h_{fi}(x|\alpha_{fi})\theta_{fi} + \sum_{i=1}^{M} h_{gi}(x|\alpha_{gi}^*)\theta_{gi}^* u - \sum_{i=1}^{M} h_{gi}(x|\alpha_{gi})\theta_{gi} u$$

$$+ \sum_{i=1}^{M} h_{gi}(x|\alpha_{gi})\theta_{gi}^* u - \sum_{i=1}^{M} h_{gi}(x|\alpha_{gi})\theta_{gi}^* u - K \, sign(s), \qquad (6.112)$$

$$\dot{s} = \varepsilon_f + \varepsilon_g u + d(t) + \sum_{i=1}^{M} \tilde{h}_{fi}\theta_{fi}^* + \sum_{i=1}^{M} h_{fi}(x|\alpha_{fi})\tilde{\theta}_{fi}$$

$$+ \sum_{i=1}^{M} \tilde{h}_{gi}\theta_{gi}^* u + \sum_{i=1}^{M} h_{gi}(x|\alpha_{gi})\tilde{\theta}_{gi} u - K \, sign(s),$$

where $\tilde{h}_{fi}$, $\tilde{h}_{gi}$, $\tilde{\theta}_{fi}$, and $\tilde{\theta}_{gi}$ are defined as follows:

$$\tilde{\theta}_{fi} = \theta_{fi}^* - \theta_{fi}$$
$$\tilde{\theta}_{gi} = \theta_{gi}^* - \theta_{gi}$$
$$\tilde{h}_{fi} = h_{fi}(x|\alpha_{fi}^*) - h_{fi}(x|\alpha_{fi})$$
$$\tilde{h}_{gi} = h_{gi}(x|\alpha_{gi}^*) - h_{gi}(x|\alpha_{gi}). \qquad (6.113)$$

The time derivative of the sliding surface can be further manipulated as follows:

$$\dot{s} = \varepsilon_f + \varepsilon_g u + d(t) + \sum_{i=1}^{M} \tilde{h}_{fi}\tilde{\theta}_{fi} + \sum_{i=1}^{M} \tilde{h}_{fi}\theta_{fi} + \sum_{i=1}^{M} h_{fi}(x|\alpha_{fi})\tilde{\theta}_{fi}$$

$$+ \sum_{i=1}^{M} \tilde{h}_{gi}\theta_{gi}^* u + \sum_{i=1}^{M} h_{gi}(x|\alpha_{gi})\tilde{\theta}_{gi} u - K \, sign(s).$$

Using the Taylor expansion of $h_{fi}(x)$ around α_{fi} and $h_{gi}(x)$ around α_{gi}, the following equation for the time derivative of the sliding surface is obtained:

$$\dot{s} = \varepsilon_f + \varepsilon_g u + d(t) + \sum_{i=1}^{M} \tilde{h}_{fi}\tilde{\theta}_{fi} + \sum_{i=1}^{M} \left(\tilde{\alpha}_{fi}^T \left. \frac{\partial h_{fi}(x)}{\partial \alpha_{fi}} \right|_{\alpha_{fi}=\alpha_{fi}} + R_f \right) \theta_{fi} + \sum_{i=1}^{M} h_{fi}(x|\alpha_{fi})\tilde{\theta}_{fi}$$

$$+ \sum_{i=1}^{M} \left(\tilde{\alpha}_{gi}^T \left. \frac{\partial h_{gi}(x)}{\partial \alpha_{gi}} \right|_{\alpha_{gi}=\alpha_{gi}} + R_g \right) \theta_{gi}^* u + \sum_{i=1}^{M} h_{gi}(x|\alpha_{gi})\tilde{\theta}_{gi} u - K \, sign(s).$$

In order to analyze the stability of the adaptation laws and control signal, the following Lyapunov function is considered.

$$V = \frac{1}{2}s^2 + \frac{1}{2\gamma_1}\sum_{i=1}^{M}\tilde{\theta}_{fi}^2 + \frac{1}{2\gamma_2}\sum_{i=1}^{M}\tilde{\alpha}_{fi}^T\tilde{\alpha}_{fi} + \frac{1}{2\gamma_3}\sum_{i=1}^{M}\tilde{\theta}_{gi}^2 + \frac{1}{2\gamma_4}\sum_{i=1}^{M}\theta_{gi}^*\tilde{\alpha}_{gi}^T\tilde{\alpha}_{gi}.$$

$$(6.114)$$

The time derivative of the Lyapunov function (6.114) is obtained as follows:

$$\dot{V} = \frac{1}{2}s\dot{s} + \frac{1}{\gamma_1}\sum_{i=1}^{M}\tilde{\theta}_{fi}\dot{\tilde{\theta}}_{fi} + \frac{1}{\gamma}\sum_{i=1}^{M}\tilde{\alpha}_{fi}^T\dot{\tilde{\alpha}}_{fi} + \frac{1}{\gamma_3}\sum_{i=1}^{M}\tilde{\theta}_{gi}\dot{\tilde{\theta}}_{gi} + \frac{1}{\gamma_4}\sum_{i=1}^{M}\theta_{gi}^*\tilde{\alpha}_{gi}^T\dot{\tilde{\alpha}}_{gi}. \quad (6.115)$$

By replacing the time derivative of the sliding surface (6.114) in (6.115), the following equation for the time derivative of error is obtained:

$$\dot{V} = \frac{1}{2}s\left(\varepsilon_f + \varepsilon_g u + d(t) + \sum_{i=1}^{M}\tilde{h}_{fi}\tilde{\theta}_{fi} + \sum_{i=1}^{M}\left(\tilde{\alpha}_{fi}^T\left.\frac{\partial h_{fi}(x)}{\partial\alpha_{fi}}\right|_{\alpha_{fi}=\alpha_{fi}} + R_f\right)\theta_{fi}\right.$$

$$+ \sum_{i=1}^{M}h_{fi}(x|\alpha_{fi})\tilde{\theta}_{fi} + \sum_{i=1}^{M}\left(\tilde{\alpha}_{gi}^T\left.\frac{\partial h_{gi}(x)}{\partial\alpha_{gi}}\right|_{\alpha_{gi}=\alpha_{gi}} + R_g\right)\theta_{gi}^*u + \sum_{i=1}^{M}h_{gi}(x|\alpha_{gi})\tilde{\theta}_{gi}u$$

$$\left.- K\,sign(s)\right) + \frac{1}{\gamma_1}\sum_{i=1}^{M}\tilde{\theta}_{fi}\dot{\tilde{\theta}}_{fi} + \frac{1}{\gamma}\sum_{i=1}^{M}\tilde{\alpha}_{fi}^T\dot{\tilde{\alpha}}_{fi} + \frac{1}{\gamma_3}\sum_{i=1}^{M}\tilde{\theta}_{gi}\dot{\tilde{\theta}}_{gi} + \frac{1}{\gamma_4}\sum_{i=1}^{M}\theta_{gi}^*\tilde{\alpha}_{gi}^T\dot{\tilde{\alpha}}_{gi}.$$

$$(6.116)$$

The following adaptation laws can be derived from (6.116):

$$\dot{\theta}_{fi} = -\dot{\tilde{\theta}}_{fi} = \gamma_1 h_{fi}(x|\alpha_{fi})s$$

$$\dot{\alpha}_{fi} = -\dot{\tilde{\alpha}}_{fi} = \gamma_2\theta_{fi}\frac{\partial h_{fi}(x)}{\partial\alpha_{fi}(x)}s$$

$$\dot{\theta}_{gi} = -\dot{\tilde{\theta}}_{gi} = \gamma_3 h_{gi}(x|\alpha_{gi})s\,u$$

$$\dot{\alpha}_{gi} = -\dot{\tilde{\alpha}}_{gi} = \gamma_4\theta_{gi}\frac{\partial h_{gi}(x)}{\partial\alpha_{gi}(x)}su. \quad (6.117)$$

By plugging in the adaptation laws of (6.117) into (6.116), the following equation for the time derivative of the Lyapunov function is obtained:

$$\dot{V} = \frac{1}{2}s\left(\varepsilon_f + \varepsilon_g u + d(t) + \sum_{i=1}^{M}\tilde{h}_{fi}\tilde{\theta}_{fi} + \sum_{i=1}^{M}R_{fi}\theta_{fi} + \sum_{i=1}^{M}R_{gi}\theta_{gi}^*u\right.$$

$$\left.- K\,sign(s)\right). \quad (6.118)$$

The remaining uncertainties can be compensated for using a large enough value for the parameter K. It is to be noted that the terms R_{fi} and R_{gi} are higher-order terms in the Taylor expansion for h_{fi} and h_{gi}, which can be considered to be sufficiently small. The remaining terms have upper bounds and can easily be compensated for using the gain K.

6.1.6   Terminal Sliding-Mode Adaptive Fuzzy Controller

The main feature of terminal sliding-mode control is that other than the sliding surface, the error of the states of the system converge to zero in finite time. In order to design a terminal sliding-mode fuzzy controller, a second-order nonlinear dynamic system is considered as follows:

$$\dot{x}_1 = x_2$$
$$\dot{x}_2 = f(x) + g(x)u, \tag{6.119}$$

where $f(x) : R^n \rightarrow R$ and $g(x) : R^n \rightarrow R$ are nonlinear functions. The tracking error for the system is represented by e, upon which the sliding surface is defined as follows:

$$s = \dot{e} + \lambda e^{q/p}. \tag{6.120}$$

In order to guarantee the finite-time convergence of the sliding manifold to zero, it is required that the following condition holds:

$$s\dot{s} < -\eta|s|. \tag{6.121}$$

The following fuzzy identifiers are designed to estimate the dynamics of the system:

$$\hat{f}(x|\hat{\theta}_f) = \sum_{i=1}^{M} \hat{\theta}_{fi} h_{fi}$$

$$\hat{g}(x|\hat{\theta}_g) = \sum_{i=1}^{M} \hat{\theta}_{gi} h_{gi}. \tag{6.122}$$

Using these fuzzy systems, the control signal of the system is designed as follows:

$$u = -\frac{1}{\sum_{i=1}^{M} \hat{\theta}_{gi} h_{gi}} \left(\sum_{i=1}^{M} \hat{\theta}_{fi} h_{fi} - \ddot{x}_d + \frac{\lambda q}{p} e^{q/p-1} \dot{e} + K sign(s) \right). \tag{6.123}$$

Considering the sliding surface of (6.120), the time derivative of the sliding surface is obtained as follows:

$$\dot{s} = \ddot{e} + \frac{q}{p}\lambda e^{q/p-1}\dot{e}$$

$$= f(x) + g(x)u - \ddot{x}_d + \frac{q}{p}\lambda e^{q/p-1}\dot{e}$$

$$= \sum_{i=1}^{M}\theta_{fi}^{*}h_{fi} + \sum_{i=1}^{M}\theta_{gi}^{*}h_{gi}u - \ddot{x}_d + \frac{q}{p}\lambda e^{q/p-1}\dot{e}, \qquad (6.124)$$

where $\sum_{i=1}^{M}\theta_{fi}^{*}h_{fi}$ and $\sum_{i=1}^{M}\theta_{gi}^{*}h_{gi}$ are the optimal fuzzy systems, which represent the nonlinear functions $f(x)$ and $g(x)$, respectively. Considering the control signal of (6.123), the time derivative of the sliding surface is obtained as follows:

$$\dot{s} = \sum_{i=1}^{M}\tilde{\theta}_{fi}h_{fi} + \sum_{i=1}^{M}\tilde{\theta}_{gi}h_{gi}u - Ksign(s). \qquad (6.125)$$

The stability analysis of the proposed controller is done using the following Lyapunov function.

$$V = \frac{1}{2}s^2 + \frac{1}{2\gamma_1}\sum_{i=1}^{M}\tilde{\theta}_{fi}^{2} + \frac{1}{2\gamma_2}\sum_{i=1}^{M}\tilde{\theta}_{gi}^{2}. \qquad (6.126)$$

The time derivative of such a Lyapunov function is obtained as follows:

$$\dot{V} = s\dot{s} + \frac{1}{\gamma_1}\sum_{i=1}^{M}\tilde{\theta}_{fi}\dot{\tilde{\theta}}_{fi} + \frac{1}{\gamma_2}\sum_{i=1}^{M}\tilde{\theta}_{gi}\dot{\tilde{\theta}}_{gi}. \qquad (6.127)$$

Considering the adaptation laws of the system as follows:

$$\dot{\tilde{\theta}}_{fi} = \gamma_1 s h_{fi}$$

$$\dot{\tilde{\theta}}_{gi} = \gamma_1 s h_{gi}u \qquad (6.128)$$

and selecting the parameter K as

$$2\eta < K. \qquad (6.129)$$

The following inequality holds for the time derivative of the Lyapunov function.

$$\dot{V} < -\eta|s|. \qquad (6.130)$$

The time taken for the sliding surface to reach *zero* is obtained as follows:

$$t_s < \frac{s(0)}{\eta}. \tag{6.131}$$

The time which takes for the error of the state to become *zero* is obtained as follows. After the time t_s, the sliding variable is equal to *zero*; hence, the following inequality holds.

$$s = 0 \rightarrow \dot{e} < -\lambda e^{q/p} \tag{6.132}$$

and further, the following is obtained:

$$e^{-\frac{q}{p}} de < -\lambda dt$$
$$\frac{p}{p-q} e^{1-\frac{q}{p}} \Big|_{e(0)}^{0} = -\lambda t \Big|_0^{t_f} \tag{6.133}$$

which finally results in the following upper bound for the time it takes for the error signal to converge to *zero*.

$$t_f < \frac{p}{p-q} e(0)^{1-q/p}. \tag{6.134}$$

Hence, the final time at which the error is guaranteed to become *zero* is obtained as equal to $t_f + t_s$.

6.2 Interval Type-2 Fuzzy Control

The degree of membership in a type-1 fuzzy structure has no uncertainty. However, there rarely exists a group of experts who all agree about the membership grades for a single input value. Different membership functions construct a range of membership degrees rather than a crisp number. Moreover, a secondary membership function is associated with the interval membership grade. It is an interval type-2 fuzzy system; if the values of all secondary membership are limited to *zero* and *one* and if they are chosen from the interval of [0, 1], the fuzzy system is called general type-2 fuzzy system. In this part, in order to be able to deal with uncertainty in a more effective way, interval type-2 fuzzy structure is used.

6.2.1 Indirect Case

Similar to the type-1 case in Sect. 6.1.2, in which the sliding-mode controller is developed for the case when the coefficient of the control signal is a function of the states of the system rather than a constant value, the interval type-2 case is considered in this section. Similarly, the nonlinear model that is used is of the following form:

$$x^{(n)} = f(x, t) + g(x, t)u + d(t)$$
$$y = x, \tag{6.135}$$

where $f(x) : R^n \rightarrow R$ and $g(x) : R^n \rightarrow R$ are unknown smooth nonlinear functions of the states of the system, which are to be approximated using interval type-2 fuzzy system, $d(t)$ is an unknown external disturbance that satisfies $|d(t)| \leq D$, and y is the measured value of the system. It is also assumed that $g(x) > 0$. The reference trajectory $r(t)$ is a sufficiently smooth desired trajectory of the system. The sliding surface is defined as follows:

$$s(t) = e^{(n-1)}(t) + \lambda_{n-1}e^{(n-2)}(t) + \cdots + \lambda_2 \dot{e}(t) + \lambda_1 e(t), \tag{6.136}$$

where $e(t)$ is the error signal, which is defined as $e(t) = x(t) - r(t)$. The sliding-mode control signal for the system is considered to be as follows:

$$u(t) = \frac{1}{\hat{g}(x|\theta_g)} \left[-\hat{f}(x|\underline{\theta}_f, \overline{\theta}_f) + \sum_{i=1}^{n-1} \lambda_i e^{(i)} - r^{(n)} - K sign(s) \right], \tag{6.137}$$

where $\hat{g}(x|\theta_g)$ and $\hat{f}(x|\underline{\theta}_f, \overline{\theta}_f)$ are Mamdani-type fuzzy identifier which approximate $g(x)$ and $f(x)$, respectively. These fuzzy systems are defined as follows:

$$\hat{f}(x|\underline{\theta}_f, \overline{\theta}_f) = 0.5 \sum_{i=1}^{M} \overline{h}_{fi}\overline{\theta}_{fi} + 0.5 \sum_{i=1}^{M} \underline{h}_{fi}\underline{\theta}_{fi}, \tag{6.138}$$

$$\hat{g}(x|\theta_g) = 0.5 \sum_{i=1}^{M} \overline{h}_{gi}\overline{\theta}_{gi} + 0.5 \sum_{i=1}^{M} \underline{h}_{gi}\underline{\theta}_{gi}, \tag{6.139}$$

where $\underline{h}_{fi}, \overline{h}_{fi}, \underline{h}_{gi}$, and $\overline{h}_{gi}$ are defined as follows:

$$\underline{h}_{fi} = \frac{\overline{w}_{fi} + \underline{w}_{fi} - sign(\underline{m}_{fi})\Delta w_{fi}}{\sum_{i=1}^{M}\left(\overline{w}_{fi} + \underline{w}_{fi}\right) - sign(\underline{m}_{fi})\Delta w_{fi}}$$

$$\overline{h}_{fi} = \frac{\overline{w}_{fi} + \underline{w}_{fi} + sign(\overline{m}_{fi})\Delta w_{fi}}{\sum_{i=1}^{M}\left(\overline{w}_{fi} + \underline{w}_{fi}\right) - sign(\underline{m}_{fi})\Delta w_{fi}}$$

$$\underline{h}_{gi} = \frac{\overline{w}_{gi} + \underline{w}_{gi} - sign(\underline{m}_{gi})\Delta w_{gi}}{\sum_{i=1}^{M}\left(\overline{w}_{gi} + \underline{w}_{gi}\right) - sign(\underline{m}_{gi})\Delta w_{gi}}$$

$$\overline{h}_{gi} = \frac{\overline{w}_{gi} + \underline{w}_{gi} + sign(\overline{m}_{gi})\Delta w_{gi}}{\sum_{i=1}^{M}\left(\overline{w}_{gi} + \underline{w}_{gi}\right) - sign(\underline{m}_{gi})\Delta w_{gi}},$$

$$(6.140)$$

where $\underline{w}_{fi}$ and $\overline{w}_{fi}$ are the lower and the upper values of the multiplication of the interval type-2 fuzzy membership functions used to approximate the function $f(X)$ and $\underline{w}_{gi}$ and $\overline{w}_{gi}$ are the lower and the upper limits of the multiplication of intervals corresponding to the membership functions of the fuzzy system $\hat{g}(x|\overline{\theta}_g, \underline{\theta}_g)$.

The control signal considered in this case is modified as follows:

$$u(t) = \frac{1}{\hat{g}(x|\theta_g)}\left[-\hat{f}(x|\underline{\theta}_f, \overline{\theta}_f) + \sum_{i=1}^{n-1}\lambda_i e^{(i)} - r^{(n)} - Ksgn(s) \right]. \qquad (6.141)$$

Theorem 6.5 *Let the nth-order nonlinear dynamic system be in the form of (6.135) with reference signal $r(t)$ and the sliding manifold as defined in (6.136). If the control signal defined in the form of (6.141) with the fuzzy approximators $\hat{f}(x|\underline{\theta}_f, \overline{\theta}_f)$ and $g(x|\theta_g)$ being defined as in (6.138) and (6.139) and the adaptation laws being as in (6.142)–(6.145) and the adaptation law of the parameter K being defined as in (6.146), then the states of system converge to the sliding manifold in finite time. Hence, the tracking error of the system will asymptotically converge to zero with the behavior that is defined in the siding motion. Furthermore, the values of the parameters of the fuzzy system remain bounded.*

$$\dot{\overline{\theta}}_{fi} = \gamma_1 \overline{h}_{fi}\, s, \qquad (6.142)$$

$$\dot{\underline{\theta}}_{fi} = \gamma_1 \underline{h}_{fi}\, s, \qquad (6.143)$$

$$\dot{\overline{\theta}}_{gi} = \gamma_2 \overline{h}_{gi}\, s\, u, \qquad (6.144)$$

$$\dot{\underline{\theta}}_{gi} = \gamma_2 \underline{h}_{gi}\, s\, u, \qquad (6.145)$$

$$\dot{K}_1 = \gamma_3|s|. \qquad (6.146)$$

Proof The adaptive parameters of the unknown T1FLSs $\hat{f}(x|\theta_f)$ and $\hat{g}(x|\theta_g)$ are $\underline{\theta}_{fi}^*$, $\overline{\theta}_{fi}^*$, $\underline{\theta}_{gi}^*$, and $\overline{\theta}_{gi}^*$, which are optimal parameters of the following optimization problems.

$$\theta_{fi}^* = arg \min_{\underline{\theta}_f, \overline{\theta}_f \in S_f} \left[\sup_{x \in R^n} |f(x,t) - \hat{f}(x|\underline{\theta}_f, \overline{\theta}_f)| \right], \tag{6.147}$$

$$\theta_{gi}^* = arg \min_{\underline{\theta}_g, \overline{\theta}_g \in S_g} \left[\sup_{x \in R^n} |g(x,t) - \hat{g}(x|\underline{\theta}_g, \overline{\theta}_g)| \right], \tag{6.148}$$

where S_f and S_g are convex sets that include $\underline{\theta}_{fi}^*$, $\overline{\theta}_{fi}^*$, $\underline{\theta}_{gi}^*$, and $\overline{\theta}_{gi}^*$. The MFAE normally exists and is defined as follows:

$$\varepsilon_f = f(x) - \hat{f}(x|\underline{\theta}_f^*, \overline{\theta}_f^*), \tag{6.149}$$

$$\varepsilon_g = g(x) - \hat{g}(x|\underline{\theta}_g^*, \overline{\theta}_g^*). \tag{6.150}$$

From the general function approximation theory, it follows that the MFAEs satisfy the following equations:

$$|\varepsilon_f| < M_f$$
$$|\varepsilon_g| < M_g ,$$

where M_f and M_g are proven to have finite constant values. Furthermore, it is possible to use a higher number of rules for the fuzzy system to make the MFAE as small as desired to obtain superior approximation. The time derivative of the sliding manifold (6.136) is obtained as follows:

$$\dot{s} = f(x) + g(x)u(t) + d(t) - r^{(n)} + \sum_{i=1}^{n-1} \lambda_i e^{(i)}. \tag{6.151}$$

It follows from (6.141) that

$$\hat{g}(x|\underline{\theta}_g, \overline{\theta}_g)u(t) + \hat{f}(x|\underline{\theta}_f, \overline{\theta}_f) - \sum_{i=1}^{n-1} \lambda_i e^{(i)} + r^{(n)} + K sign(s) = 0. \tag{6.152}$$

By adding (6.152)–(6.141), the following equation is obtained:

$$\dot{s} = f(x) + g(x)u + d(t) - 0.5\sum_{i=1}^{M}\underline{h}_{fi}\underline{\theta}_{fi} - 0.5\sum_{i=1}^{M}\overline{h}_{fi}\overline{\theta}_{fi} - 0.5\sum_{i=1}^{M}\underline{h}_{gi}\underline{\theta}_{gi} - 0.5\sum_{i=1}^{M}\overline{h}_{gi}\overline{\theta}_{gi} - Ksign(s)$$

$$= f(x) - \hat{f}(x|\underline{\theta}_f^*, \overline{\theta}_f^*) - \hat{g}(x|\underline{\theta}_g^*, \overline{\theta}_g^*)u + g(x)u + d(t) + \hat{f}(x|\underline{\theta}_f^*, \overline{\theta}_f^*) - 0.5\sum_{i=1}^{M}\underline{h}_{fi}\underline{\theta}_{fi} - 0.5\sum_{i=1}^{M}\overline{h}_{fi}\overline{\theta}_{fi}$$

$$+ \hat{g}(x|\underline{\theta}_g^*, \overline{\theta}_g^*)u - 0.5\sum_{i=1}^{M}\underline{h}_{gi}\underline{\theta}_{gi}u - 0.5\sum_{i=1}^{M}\overline{h}_{gi}\overline{\theta}_{gi}u - Ksign(s)$$

$$= f(x) - \sum_{i=1}^{M}h_{fi}\theta_{fi}^* + g(x)u - \sum_{i=1}^{M}h_{gi}\theta_{gi}^*u + d(t) + \sum_{i=1}^{M}h_{fi}\theta_{fi}^* - \sum_{i=1}^{M}h_{fi}\theta_{fi}$$

$$+ \sum_{i=1}^{M}h_{gi}\theta_{gi}^*u - \sum_{i=1}^{M}h_{gi}\theta_{gi}u - Ksign(s)$$

$$= \varepsilon_f + \varepsilon_g u + d(t) + 0.5\sum_{i=1}^{M}\underline{h}_{fi}\tilde{\underline{\theta}}_{fi} + 0.5\sum_{i=1}^{M}\overline{h}_{fi}\tilde{\overline{\theta}}_{fi} + 0.5\sum_{i=1}^{M}\underline{h}_{fi}\tilde{\underline{\theta}}_{fi} + \sum_{i=1}^{M}h_{gi}\tilde{\overline{\theta}}_{gi}u - Ksign(s),$$

where $\tilde{\underline{\theta}}_{fi}$, $\tilde{\overline{\theta}}_{fi}$, $\tilde{\underline{\theta}}_{gi}$, and $\tilde{\overline{\theta}}_{gi}$ are defined as follows:

$$\tilde{\underline{\theta}}_{fi} = \underline{\theta}_{fi}^* - \underline{\theta}_{fi}$$

$$\tilde{\overline{\theta}}_{fi} = \overline{\theta}_{fi}^* - \overline{\theta}_{fi}$$

$$\tilde{\underline{\theta}}_{gi} = \underline{\theta}_{gi}^* - \underline{\theta}_{gi}$$

$$\tilde{\overline{\theta}}_{gi} = \overline{\theta}_{gi}^* - \overline{\theta}_{gi}. \tag{6.153}$$

In order to analyze the stability of the system, the following Lyapunov function is considered.

$$V = \frac{1}{2}s^2 + \frac{1}{2\gamma_1}\sum_{i=1}^{M}\tilde{\underline{\theta}}_{fi}^2 + \frac{1}{2\gamma_2}\sum_{i=1}^{M}\tilde{\overline{\theta}}_{fi}^2 + \frac{1}{2\gamma_3}\sum_{i=1}^{M}\tilde{\underline{\theta}}_{gi}^2 + \frac{1}{2\gamma_4}\sum_{i=1}^{M}\tilde{\overline{\theta}}_{gi}^2 + \frac{1}{2\gamma_5}\tilde{K}_1^2,$$

$$\tag{6.154}$$

where $0 < \gamma_1, \gamma_2, \gamma_3, \gamma_4$, and γ_5 are the learning rates of adaptive parameters. Theoretically, they can be chosen as being equal to any positive value. Moreover, the parameter $\tilde{K}_1$ is the error of parameters and are defined as being equal to $\tilde{K}_1 = K_1 - K_1^*$ with K_1^* being the unknown optimal value of the parameter. The time derivative of this Lyapunov function (6.154) is obtained as follows:

$$\dot{V} = \sum_{i=1}^{M}\tilde{\underline{\theta}}_{fi}(\underline{h}_{fi}s + \gamma_1^{-1}\dot{\underline{\theta}}_{fi}) + \sum_{i=1}^{M}\tilde{\overline{\theta}}_{fi}(\overline{h}_{fi}s + \gamma_2^{-1}\dot{\overline{\theta}}_{fi})$$

$$+ \sum_{i=1}^{M}\tilde{\underline{\theta}}_{gi}(\underline{h}_{gi}su + \gamma_3^{-1}\dot{\underline{\theta}}_{gi}) + \sum_{i=1}^{M}\tilde{\overline{\theta}}_{gi}(\overline{h}_{gi}su + \gamma_4^{-1}\dot{\overline{\theta}}_{gi})$$

$$+ \frac{1}{\gamma_5} \tilde{K}_1 \dot{K}_1 + s\left(\varepsilon_f + \varepsilon_g u + d(t)\right) - K_1|s|. \tag{6.155}$$

Considering the adaptation laws of (6.142) and (6.146), one obtains the following equation for the time derivative of the Lyapunov function.

$$\dot{V} = s\left(\varepsilon_f + \varepsilon_g u + d(t)\right) - K_1^*|s|. \tag{6.156}$$

Taking $K_1^* > M_f + M_g u + D + \eta$, the following equation is obtained.

$$\dot{V} < -\eta|s|. \tag{6.157}$$

This concludes the proof and guarantees that the sliding manifold converges to *zero* asymptatically. Moreover, the adaptive parameters of the system remain bounded.

6.2.2   Direct Controller

Direct adaptive interval type-2 fuzzy controller is designed in this section. The non-linear dynamic system is considered to be of the following form.

$$x^{(n)} = f(x) + g(x)u$$
$$y = x, \tag{6.158}$$

where $f(x) : R^n \rightarrow R$ and $g(x) : R^n \rightarrow R$ are sufficiently smooth nonlinear functions. It is further assumed that $0 < g_l < g(x) < g_u$ and $|\dot{g}(x)| < M_{\dot{g}}$. Let the tracking error be

$$e = x - r = (x - r, \dot{x} - \dot{r}, \dots, x^{(n-1)} - r^{(n-1)}) = (e, \dot{e}, \dots, e^{(n-1)}). \tag{6.159}$$

Then, a sliding surface is defined based on these error signals as follows:

$$s = e^{(n-1)} + \lambda_{n-1} e^{(n-2)} + \cdots + \lambda_2 \dot{e} + \lambda_1 e, \tag{6.160}$$

where the parameters $\lambda_1, \dots, \lambda_{n-1}$ are chosen such that the sliding surface is stable. Considering (6.158), we have the following equation:

$$\dot{s} = f(x) + g(x)u + d(t) - r^{(n)} + \lambda_{n-1} e^{(n-1)} + \cdots + \lambda_2 \ddot{e} + \lambda_1 \dot{e}. \tag{6.161}$$

The sliding-mode control signal to guarantee the stability of the system in this case is as follows:

$$u^* = \frac{-f(x)}{g(x)} - \frac{\rho(t)}{g(x)} = p(x) + q(x)\rho(t), \tag{6.162}$$

where $\rho(t)$ is defined as follows:

$$\rho(t) = -r^{(n)} + \lambda_{n-1}e^{(n-1)} + \cdots + \lambda_2\ddot{e} + \lambda_1\dot{e} + K_1 sign(s) + K_2 s. \tag{6.163}$$

Two separate fuzzy systems are considered to approximate $p(x)$ and $q(x)$; these fuzzy systems are considered to be as follows:

$$\hat{p}(x|\hat{\overline{\theta}}_p, \hat{\underline{\theta}}_p) = 0.5 \sum_{i=1}^{M} \underline{h}_{pi}\hat{\underline{\theta}}_{pi} + 0.5 \sum_{i=1}^{M} \overline{h}_{pi}\hat{\overline{\theta}}_{pi}$$

$$\hat{q}(x|\hat{\overline{\theta}}_q, \hat{\underline{\theta}}_q) = 0.5 \sum_{i=1}^{M} \underline{h}_{qi}\hat{\underline{\theta}}_{qi} + 0.5 \sum_{i=1}^{M} \overline{h}_{qi}\hat{\overline{\theta}}_{qi}. \tag{6.164}$$

Considering (6.163) and (6.164), it follows that

$$f(x) + g(x)\left(\hat{p}(x|\overline{\theta}_p^*, \underline{\theta}_p^*) + \hat{q}(x|\overline{\theta}_q^*, \underline{\theta}_q^*)\rho(x)\right) - r^{(n)} + \lambda_{n-1}e^{(n-1)} + \cdots + \lambda_2\ddot{e} + \lambda_1\dot{e} + K_1 sign(s) + K_2 s = 0. \tag{6.165}$$

Furthermore, by adding and subtracting the term $g(x)\left(\hat{p}(x|\hat{\overline{\theta}}_p, \hat{\underline{\theta}}_p) + \hat{q}(x|\hat{\overline{\theta}}_q, \hat{\underline{\theta}}_q)\rho(x)\right)$, the following equation is obtained.

$$f(x) + g(x)\left(0.5 \sum_{i=1}^{M} \underline{h}_{pi}\tilde{\underline{\theta}}_{pi} + 0.5 \sum_{i=1}^{M} \overline{h}_{pi}\tilde{\overline{\theta}}_{pi} + 0.5 \sum_{i=1}^{M} \underline{h}_{qi}\tilde{\underline{\theta}}_{qi}\rho(x) + 0.5 \sum_{i=1}^{M} \overline{h}_{qi}\tilde{\overline{\theta}}_{qi}\rho(x)\right)$$

$$+ g(x)\left(\hat{p}(x|\hat{\overline{\theta}}_p, \hat{\underline{\theta}}_p) + \hat{q}(x|\hat{\overline{\theta}}_q, \hat{\underline{\theta}}_q)\rho(x)\right)$$

$$- r^{(n)} + \lambda_{n-1}e^{(n-1)} + \cdots + \lambda_2\ddot{e} + \lambda_1\dot{e} + K_1 sign(s) + K_2 s = 0. \tag{6.166}$$

By adding (6.166)–(6.160), the following equation is obtained.

$$\dot{s} = -g(x)\left(0.5 \sum_{i=1}^{M} \underline{h}_{pi}\tilde{\underline{\theta}}_{pi} + 0.5 \sum_{i=1}^{M} \overline{h}_{pi}\tilde{\overline{\theta}}_{pi} + 0.5 \sum_{i=1}^{M} \underline{h}_{qi}\tilde{\underline{\theta}}_{qi}\rho(x) + 0.5 \sum_{i=1}^{M} \overline{h}_{qi}\tilde{\overline{\theta}}_{qi}\rho(x)\right)$$

$$+ d(t) - K_1 sign(s) - K_2 s. \tag{6.167}$$

The following Lyapunov function is considered for analyzing the stability of the system:

$$V = \frac{1}{2g(x)}s^2 + \frac{1}{\gamma_1} \sum_{i=1}^{M} \tilde{\underline{\theta}}_{pi}^T \tilde{\underline{\theta}}_{pi} + \frac{1}{\gamma_2} \sum_{i=1}^{M} \tilde{\overline{\theta}}_{pi}^T \tilde{\overline{\theta}}_{pi} + \frac{1}{\gamma_3} \sum_{i=1}^{M} \tilde{\underline{\theta}}_{qi}^T \tilde{\underline{\theta}}_{qi} + \frac{1}{\gamma_4} \sum_{i=1}^{M} \tilde{\overline{\theta}}_{qi}^T \tilde{\overline{\theta}}_{qi}. \tag{6.168}$$

The time derivative of this Lyapunov function is obtained as follows:

$$
\dot{V} = -\frac{\partial g(x)}{\partial x^{(n-1)}} x^{(n)} \frac{1}{2g^2(x)} s^2 + \frac{1}{g(x)} s \left(-g(x) \left(\sum_{i=1}^{M} \underline{h}_{pi} \tilde{\underline{\theta}}_{pi} + \sum_{i=1}^{M} \overline{h}_{pi} \tilde{\overline{\theta}}_{pi} \right.\right.
$$

$$
+ \sum_{i=1}^{M} \underline{h}_{qi} \tilde{\underline{\theta}}_{qi} \rho(x) + \sum_{i=1}^{M} \overline{h}_{qi} \tilde{\overline{\theta}}_{qi} \rho(x)) + 2d(t) - 2K_1 sign(s) - 2K_2 s \bigg)
$$

$$
+ \frac{1}{\gamma_1} \sum_{i=1}^{M} \tilde{\underline{\theta}}_{pi}^T \dot{\tilde{\underline{\theta}}}_{pi} + \frac{1}{\gamma_2} \sum_{i=1}^{M} \tilde{\overline{\theta}}_{pi}^T \dot{\tilde{\overline{\theta}}}_{pi} + \frac{1}{\gamma_3} \sum_{i=1}^{M} \tilde{\underline{\theta}}_{qi}^T \dot{\tilde{\underline{\theta}}}_{qi} + \frac{1}{\gamma_4} \sum_{i=1}^{M} \tilde{\overline{\theta}}_{qi}^T \dot{\tilde{\overline{\theta}}}_{qi}. \quad (6.169)
$$

The adaptation laws of the system are considered so that the terms with unknown signs are eliminated from the time derivative of the Lyapunov function. The adaptation laws of the system are as follows:

$$
\dot{\underline{\theta}}_{pi} = \gamma_1 \underline{h}_{pi} s
$$

$$
\dot{\overline{\theta}}_{pi} = \gamma_2 \overline{h}_{pi} s
$$

$$
\dot{\underline{\theta}}_{pi} = \gamma_3 \underline{h}_{pi} s \rho
$$

$$
\dot{\overline{\theta}}_{pi} = \gamma_3 \overline{h}_{pi} s \rho. \quad (6.170)
$$

Considering (6.170), the following equation for the time derivative of the Lypunov function is obtained:

$$
\dot{V} = -\frac{\partial g(x)}{\partial x^{(n-1)}} x^{(n)} \frac{1}{2g^2(x)} s^2 + \frac{1}{g(x)} s \left(2d(t) - 2K_1 sign(s) - 2K_2 s \right).
$$

In order to guarantee the stability of the system, the parameter K_2 is selected as follows:

$$
\frac{1}{4} Sup_{x \in \Omega} \left| \frac{\partial g(x)}{\partial x^{(n-1)}} \frac{x^{(n)}}{g^2(x)} \right| < K_2, \quad (6.171)
$$

where Ω is a compact set that includes the origin. Moreover, the parameter K_1 is selected as follows:

$$
max \left\{ \eta, \frac{2D}{g_{min}} \right\} < K_1 \quad (6.172)
$$

which, in turn, guarantees that the following inequality is valid for the time derivative of error.

$$\dot{V} \le -\eta |s|.$$

This concludes the stability of the system in a compact set Ω.

6.3 Robustness Issues

Unmodeled dynamics is inevitable in modeling a nonlinear dynamic system. For instance, the behavior of a spring in most mechanical systems is considered to be linear rather than nonlinear to avoid complications in the model. Friction is another phenomenon that may be ignored during modeling processes. Time delay, back-lash, dead zone, and hysteresis are among those behaviors that normally exist in physical systems but are mostly ignored due to simplification. Other than unmodeled dynamics, disturbance, noise, and minimum functional approximation error are other sources of uncertainty in a nonlinear dynamic system. Disturbances are typically unknown signals with large signal amplitudes whose frequencies are close to the frequency of the reference signal. These sources of uncertainty may disturb the perfect mathematical stability analysis of the system and cause drifts in the parameters of the system. In other words, even though the stability analysis of the system guarantees that the parameters of the system are bounded, the uncertainties that normally exist in the system may result in divergence in few parameters, which consequently causes instability in the system. In order to guarantee the robustness of the system in the presence of unmodeled dynamics, some modifications to the adaptation laws must be introduced so as to make it possible for them to deal with these sorts of uncertainties.

6.3.1 Modification of the Adaptation Law Using a σ Term

In this case, a term is added to the adaptation law to avoid its bursting. Consider the sliding-mode controller that is designed in Sect. 6.1.1. The modification done to the adaptation law using the σ term is as follows:

$$\dot{\theta}_{fi} = \gamma_1 \, h_i \, s - \sigma \gamma_1 \theta_{fi}. \tag{6.173}$$

It can be seen from (6.173) that when θ_{fi} becomes too large, the term $-\sigma \gamma_1 \theta_{fi}$ would have a large value, which pushes the parameter θ_{fi} toward *zero* and prevents it from diverging. However, using this modification modifies the stability analysis. The time derivative of the Lyapunov function in (6.23) is modified as follows:

$$\dot{V} \le -\eta|s| + \sum_{i=1}^{M} \sigma \theta_{fi} \tilde{\theta}_{fi}$$

$$= -\eta|s| - \sum_{i=1}^{M} \sigma \left(\theta_{fi} - \frac{\theta_{fi}^{*}}{2} \right)^2 + \sum_{i=1}^{M} \frac{\sigma \theta_{fi}^{*2}}{4}$$

$$\le -\eta|s| + \sum_{i=1}^{M} \frac{\sigma \theta_{fi}^{*2}}{4}. \tag{6.174}$$

As it can be seen from the obtained inequality, the time derivative of the Lyapunov function is no more a negative one. However, the parameter σ can be selected to have a very small value to have less impact on the time derivative of the Lyapunov function. In this case, the Lyapunov function converges to a neighborhood of *zero*, in which, $|s| \le \sum_{i=1}^{M} \frac{\sigma \theta_{fi}^{*2}}{4\eta}$, and remains there.

The main drawback of modifying the adaptation law using the σ term is that it is always trying to force the parameters toward *zero*. In order to avoid this phenomenon, the ε term is introduced.

6.3.2   Modification of the Adaptation Law Using a ε Term

In this case, the term that is added to the adaptation law to avoid its divergence is modified so as not to direct the parameters toward zero. This modification is as follows:

$$\dot{\theta}_{fi} = \gamma_1 \, h_i \, s - \varepsilon \gamma_1 |s| \theta_{fi}. \tag{6.175}$$

The benefit of using the adaptation law of (6.175) is that in this case similar to the σ term when θ_{fi} becomes too large the term $-\varepsilon \gamma_1 |s| \theta_{fi}$ prevents bursting in θ_{fi}. However, when the sliding manifold is close enough to *zero*, the term $-\varepsilon \gamma_1 |s| \theta_{fi}$ is also be small enough and the whole time derivative of the parameter θ_{fi} becomes very small and adaptation stops. In this case, convergence to *zero* is avoided.

In this case, the time derivative of the Lyapunov function in (6.23) is modified as follows:

$$\dot{V} \le -\eta|s| + \sum_{i=1}^{M} \varepsilon \theta_{fi} \tilde{\theta}_{fi} |s|$$

$$= -\eta|s| - \sum_{i=1}^{M} \varepsilon |s| \left(\theta_{fi} - \frac{\theta_{fi}^{*}}{2} \right)^2 + \sum_{i=1}^{M} \frac{\varepsilon \theta_{fi}^{*2}}{4} |s|$$

$$\le -\left(\eta - \sum_{i=1}^{M} \frac{\varepsilon \theta_{fi}^{*2}}{4} \right) |s|. \tag{6.176}$$

Hence, it is possible to consider the parameters η and ε such that desired convergence speed is satisfied.

6.4 Guaranteed Cost Controller Design

In this section, optimal tuning of the parameters of the controller parameters to guarantee the least integral of squared of sliding surface is obtained.

6.4.1 Constant Control Signal Coefficient Case

Consider a nth-order nonlinear dynamic system as follows:

$$x^{(n)} = f(x, t) + u(t) \tag{6.177}$$

with the same nonlinear functions as mentioned earlier in the description of (6.1). The reference signal, tracking error, and the sliding manifold are the same as defined in Sect. 6.1.1. In this case, the time derivative of the sliding manifold is obtained as follows:

$$\dot{s} = f(x) + u(t) - r^{(n)} + \lambda_{n-1}e^{(n-1)} + \cdots + \lambda_2\ddot{e} + \lambda_1\dot{e}. \tag{6.178}$$

The control signal for this system is considered to be as follows:

$$u(t) = -\hat{f}(x|\theta_f) + r^{(n)} - \sum_{i=1}^{n-1}\lambda_i e^{(i)}, \tag{6.179}$$

where the T1FLS $\hat{f}(x|\theta_f)$ considered to estimate the nonlinear smooth function $f(x, t)$ is as follows. It is desired to optimally tune the parameters of the fuzzy system, θ_f, such that it minimizes a cons function based on the sliding manifold. The T1FLS is as follows:

$$\hat{f}(x|\theta_f) = \sum_{i=1}^{M} h_i\theta_{fi}, \tag{6.180}$$

where h_i is the normalized firing strength of ith rule, θ_i is its corresponding consequent part parameters, and M is the number of rules.

Theorem 6.6 *Consider the nth-order nonlinear dynamic system of (6.1) whose states are to track the reference signal r(t) with the sliding manifold as defined in (6.3). If the control signal (6.11) is applied with $\hat{f}(x|\theta_f)$ is as given in (6.10) whose parameters are adaptively tuned as in (6.12) and u_r is defined as equal to $-Ksgn(s)$,* *then sliding manifold converges to zero in finite time resulting in sliding-mode behavior for the system. Hence, the tracking error of the system will asymptotically converge to zero, while parameter errors remain bounded.*

$$\dot{\theta}_{fi} = \gamma_1 \, h_i \, s. \tag{6.181}$$

Proof The unknown optimal parameters of the T1FLS (θ_{fi}^*) are defined as follows:

$$\theta_{fi}^* = arg \min_{\theta_f \in S_f} \left[\sup_{x \in R^n} |f(x,t) - \hat{f}(x|\theta_f)| \right], \tag{6.182}$$

where S_f is a convex set to which the optimal values of T1FLS belong. The MFAE is ignored in this case. By substituting (6.179) in (6.178), the following equation is obtained:

$$\dot{s} = \sum_{i=1}^{M} h_i \theta_{fi}^* - \sum_{i=1}^{M} h_{fi} \theta_{fi}. \tag{6.183}$$

Both sides of Eq. (6.183) are filtered using the filter $G(p)$ as follows:

$$G(p) = \frac{\lambda}{p + \lambda}, \tag{6.184}$$

where p is the time derivative operator and λ is the pole of the filter which is assumed to have sufficiently large value. Hence, the following equation is obtained.

$$\frac{\lambda p s}{p + \lambda} = \sum_{i=1}^{M} \frac{\lambda h_{fi}}{p + \lambda} \theta_{fi}^* - \sum_{i=1}^{M} \theta_{fi} \frac{\lambda h_{fi}}{p + \lambda} + \sum_{i=1}^{M} \theta_{fi} \frac{\lambda h_{fi}}{p + \lambda} - \sum_{i=1}^{M} \frac{\lambda h_{fi} \theta_{fi}}{p + \lambda}. \tag{6.185}$$

Considering fast enough poles for the $G(p)$ filter, the following approximate equation is obtained.

$$s \approx \frac{\lambda p s}{p + \lambda} \approx \sum_{i=1}^{M} \omega_{fi} \theta_{fi}^* - \sum_{i=1}^{M} \omega_{fi} \theta_{fi}, \tag{6.186}$$

where ω_{fi} is defined as follows:

$$\omega_{fi} = \frac{\lambda h_{fi}}{p + \lambda}. \tag{6.187}$$

In order to obtain the adaptation laws of θ_{fi}'s, the following cost function is defined.

$$J = \int_0^t s^2(\tau) d\tau. \tag{6.188}$$

Considering (6.186), the following equation is obtained.

$$J = \frac{1}{2}\int_0^t \left(\sum_{i=1}^{M} \omega_{fi}\theta_{fi}^* - \sum_{i=1}^{M} \omega_{fi}\theta_{fi} \right)^2 d\tau. \tag{6.189}$$

The following definitions are made to make the equations more compact.

$$\Theta = \left[\theta_{f1}, \theta_{f2}, \ldots, \theta_{fM} \right]^T$$
$$\Theta^* = \left[\theta_{f1}^*, \theta_{f2}^*, \ldots, \theta_{fM}^* \right]^T$$
$$\Omega = \left[\omega_{f1}, \omega_{f2}, \ldots, \omega_{fM} \right]^T. \tag{6.190}$$

The cost function in the vector form is obtained as follows:

$$J = \frac{1}{2}\int_0^t \left(\Theta^{*T}\Omega(\tau) - \Theta^T(t)\Omega(\tau) \right)^2 d\tau. \tag{6.191}$$

Taking the partial derivative of the cost function with respect to Θ, the following equation is obtained.

$$\frac{\partial J}{\partial \Theta} = \int_0^t \Omega \left(\Omega^T\Theta^* - \Omega^T\Theta \right) d\tau = 0. \tag{6.192}$$

Solving (6.192) for Θ, the following equation is obtained.

$$\Theta = \left(\int_0^t \Omega\Omega^T d\tau \right)^{-1} \int_0^t \Omega\Omega^T\Theta^* d\tau. \tag{6.193}$$

Equation (6.193) is the batch solution for Θ. The following definition is made.

$$P(t) = \left(\int_0^t \Omega\Omega^T d\tau \right)^{-1}. \tag{6.194}$$

We have the following equations.

$$P^{-1}(t)P(t) = I$$
$$\frac{P^{-1}(t)}{dt}P(t) + P^{-1}(t)\frac{P(t)}{dt} = 0 \tag{6.195}$$

which further results in the following equation for the time derivative of P^{-1}.

$$\frac{dP^{-1}(t)}{dt} = -P^{-1}(t)\frac{P(t)}{dt}P^{-1}(t).$$ (6.196)

Using some simple algebraic manipulation, the following equation is obtained for the time derivative of the matrix $P(t)$.

$$\frac{dP(t)}{dt} = -P(t)\Omega\Omega^T P(t).$$ (6.197)

Equation (6.192) can be further manipulated as follows:

$$\int_0^t \Omega\left(\Omega^T \Theta^* - \Omega^T \Theta\right) d\tau = \int_0^t \Omega\Omega^T \Theta^* d\tau - P^{-1}\theta.$$ (6.198)

Taking time derivative of (6.198), the following equation is obtained.

$$\Omega\Omega^T \Theta^* - \frac{dP^{-1}}{dt}\theta - P^{-1}\dot{\Theta} = 0$$ (6.199)

which results in the following adaptation law for the parameter Θ.

$$\dot{\Theta} = P\Omega\left(\Omega^T \Theta^* - \Omega^T \Theta\right) = P\Omega s.$$ (6.200)

The following Lyapunov function is considered to prove the stability of the adaptation law.

$$V = \frac{1}{2}\tilde{\Theta}^T(t)P^{-1}(t)\tilde{\Theta}(t).$$ (6.201)

Taking the time derivative of the Lyapunov function, the following equations are obtained.

$$\begin{aligned}
\dot{V} &= \tilde{\Theta}^T(t)P^{-1}(t)\dot{\tilde{\Theta}}(t) + \frac{1}{2}\tilde{\Theta}^T(t)\frac{dP^{-1}(t)}{dt}\tilde{\Theta} \\
&= -\tilde{\Theta}P^{-1}P\Omega s + \frac{1}{2}\tilde{\Theta}^T P\Omega\Omega^T \tilde{\Theta} \\
&= -\tilde{\Theta}P^{-1}P\Omega\Omega^T \tilde{\Theta} + \frac{1}{2}\tilde{\Theta}^T P\Omega\Omega^T \tilde{\Theta} \\
&= -\frac{1}{2}\tilde{\Theta}^T P\Omega\Omega^T \tilde{\Theta} \\
&= -\frac{1}{2}s^2
\end{aligned}$$ (6.202)

which indicate the $\dot{V} \leq 0$ hence $V \in L_\infty$. Since the Lyapunov function V includes $\tilde{\Theta}$, it follows that $\tilde{\Theta} \in L_\infty$ which further results in $s \in L_\infty$. Furthermore, it can easily follow that $s \in L_2$ and $\dot{s} \in L_\infty$. Hence, according to Barbalat's lemma: $\lim_{t \to \infty} s(t) = 0$.

6.5 Type-2 Feedback Error Learning Controller

In this section, an adaptive learning algorithm is proposed for an interval type-2 fuzzy controller. The structure of the system is based on the feedback error learning method. The stability of the adaptation laws is proved using Lyapunov theory. In order to show the implementability of the proposed method, it is used to control a real-time laboratory setup 2-DOF helicopter. It is shown that the proposed controller can be implemented in a low cost embedded system and can successfully control a highly nonlinear dynamic system.

Although, the stability analysis of fuzzy controllers is a common practice nowadays, it was not addressed until early 90s. The first method which tries to add stability analysis to intelligent approaches like neural networks and T1FLCs was FEL structure. In this structure, a traditional controller works in parallel with an intelligent controller. This method was first introduced by Kawato in an effort to establish a stable controller which can learn the inverse dynamic of the system under control [5]. The basic idea is to design a robust traditional controller to stabilize the closed-loop system and then the intelligent controller learns the nonlinear inverse dynamics of the system to improve the performance of overall system. In this case, since in the very first moments of run of simulation, appropriate initial conditions of the system may not be chosen, it is required to use the conventional controller to act like a guideline to make it possible for the intelligent controller to have enough time to tune its parameters.

The FEL controller designed in this section benefits from a SMC theory-based learning-based adaptation laws. The proposed method benefits from an adaptive learning rate whose adaptation law is derived using an appropriate Lyapunov function. The proposed approach is a model free one and its superiority over model-based approaches is that they can deal with modeling uncertainties and disturbances more efficiently. The adaptation mechanism for the controller and the learning rate makes it possible to control the system with less a priori knowledge. This control method is tested on the control a real-time laboratory setup 2-DOF helicopter. The performance of the system in the presence of disturbance is also studied. It is observed that the proposed controller can control the 2-DOF helicopter in the presence of load disturbances.

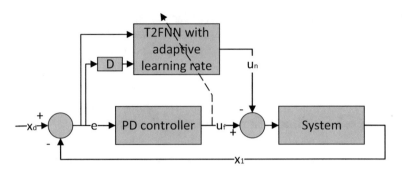

Fig. 6.4 The structure of the type-2 fuzzy FEL controller

6.6 The Proposed Controller Structure

The controller is in FEL structure which contains a proportional derivative *PD* controller and an IT2FLS in parallel as in Fig. 6.4.

6.6.1 PD Controller

The tracking error considered for the system is defined as $e = x_d - x_1$ where x_1 is state variable of the plant which is also considered to be its output. The mathematical formula for the *PD* controller is selected as to be equal to $u_f = k_D \dot{e} + k_P e$, where K_D and K_P are the gains of the conventional controller.

6.6.2 Interval Type-2 Fuzzy Logic Systems

The IT2FLS is considered to have two inputs as $e(t)$ and $De(t)$. The structure considered for the IT2FLS is a zero-order TS fuzzy model with its MFs considered being Gaussian IT2MFs with interval mean values. In this structure, IT2MFs are used in the premise part and crisp values are used for the consequent part parameters. The IF–THEN rules of the system are considered to be as follows:

$$R_{ij} : \text{If } e \text{ is } F_{1_i} \text{ and If } \dot{e} \text{ is } F_{2_j} \text{ Then } u_n = \theta_{ij},$$

where F_{1_i} and F_{2_j} are type-2 fuzzy sets for the antecedent part of the fuzzy system and θ_{ij} are the crisp numbers of the consequent part. Right-most for each rule are computed by multiplying the right-most of the MFs and the left-most of each rule are computed by multiplying the left-most of the MFs.

$$\overline{\zeta_{ij}} = \overline{\mu_{1_i}}(e)\overline{\mu_{2_j}}(\dot{e}) \text{ and } \underline{\zeta_{ij}} = \underline{\mu_{1_i}}(e)\underline{\mu_{2_j}}(\dot{e}), \tag{6.203}$$

where $\overline{\mu_{1_i}}$ and μ_{2_j} are the right-most and the left-most of the corresponding rules. The IT2MFs of the system for the inputs e and $\dot{e}$ are shown as $\mu_{1_i}(e), i = 1, \ldots, I$ and $\mu_{2_j}(\dot{e}), j = 1, \ldots, J$. The right most and left most of the MFs are defined as follows:

$$\underline{\mu_{1i}(e)} = \begin{cases} \exp\left(-(e-\overline{m_{1i}})^2/\sigma_{1i}^2\right) & , e < (\underline{m_{1i}} + \overline{m_{1i}})/2 \\ \exp\left(-(e-\underline{m_{1i}})^2/\sigma_{1i}^2\right) & , e > (\underline{m_{1i}} + \overline{m_{1i}})/2, \end{cases}$$

$$\overline{\mu_{1i}(e)} = \begin{cases} \exp\left(-(e-\underline{m_{1i}})^2/\sigma_{1i}^2\right) & , e < \underline{m_{1i}} \\ 1 & , \underline{m_{1i}} < e < \overline{m_{1i}} \\ \exp\left(-(e-\overline{m_{1i}})^2/\sigma_{1i}^2\right) & , e > \overline{m_{1i}}, \end{cases}$$

$$\underline{\mu_{2j}(\dot{e})} = \begin{cases} \exp\left(-(\dot{e}-\overline{m_{2j}})^2/\sigma_{2j}^2\right) & , \dot{e} < (\underline{m_{2j}} + \overline{m_{2j}})/2 \\ \exp\left(-(\dot{e}-\underline{m_{2j}})^2/\sigma_{2j}^2\right) & , \dot{e} > (\underline{m_{2j}} + \overline{m_{2j}})/2, \end{cases}$$

$$\overline{\mu_{2j}(\dot{e})} = \begin{cases} \exp\left(-(\dot{e}-\underline{m_{2j}})^2/\sigma_{2j}^2\right) & , \dot{e} < \underline{m_{2j}} \\ 1 & , \underline{m_{2j}} < \dot{e} < \overline{m_{2j}} \\ \exp\left(-(\dot{e}-\overline{m_{2j}})^2/\sigma_{2j}^2\right) & , \dot{e} > \overline{m_{2j}}. \end{cases}$$

The output of the IT2FLS is obtained as follows:

$$u_n = q \sum_{i=1}^{I} \sum_{j=1}^{J} \theta_{ij} \underline{\tilde{\zeta}_{ij}} + (1-q) \sum_{i=1}^{I} \sum_{j=1}^{J} \theta_{ij} \overline{\tilde{\zeta}_{ij}}, \tag{6.204}$$

where the parameter q adjusts the contribution of left-most and right-most in the output and $\overline{\tilde{\zeta}_{ij}}$ are the normalized firing strengths, which are calculated as

$$\overline{\tilde{\zeta}_{ij}} = \frac{\overline{\zeta_{ij}}}{\sum_{i=1}^{I}\sum_{j=1}^{J}\overline{\zeta_{ij}}} , \underline{\tilde{\zeta}_{ij}} = \frac{\underline{\zeta_{ij}}}{\sum_{i=1}^{I}\sum_{j=1}^{J}\underline{\zeta_{ij}}}$$

thus

$$0 < \overline{\tilde{\zeta}_{ij}} < 1 , 0 < \underline{\tilde{\zeta}_{ij}} < 1$$

$$\sum_{i=1}^{I}\sum_{j=1}^{J}\overline{\tilde{\zeta}_{ij}} = 1 , \sum_{i=1}^{I}\sum_{j=1}^{J}\underline{\tilde{\zeta}_{ij}} = 1.$$

$\overline{\tilde{\zeta}_{ij}}$ and $\underline{\tilde{\zeta}_{ij}}$ can be written as follows:

$$\underline{\tilde{\zeta}_{ij}} = \underline{\zeta}_{ij} + \Delta\underline{\tilde{\zeta}_{ij}} \,, \overline{\tilde{\zeta}_{ij}} = \overline{\zeta}_{ij} + \Delta\overline{\tilde{\zeta}_{ij}}, \tag{6.205}$$

where $\tilde{\zeta}_{ij}$ has a value which falls in the interval of left most and right most of the normalizes values of the rules. Using (6.205) the output can be rewritten as

$$u_n = \sum_{i=1}^{I}\sum_{j=1}^{J}\theta_{ij}\tilde{\zeta}_{ij} + q\sum_{i=1}^{I}\sum_{j=1}^{J}\theta_{ij}\Delta\underline{\tilde{\zeta}_{ij}} + (1-q)\sum_{i=1}^{I}\sum_{j=1}^{J}\theta_{ij}\Delta\overline{\tilde{\zeta}_{ij}}.$$

Finally, the control signal is defined as $u_c = u_f - u_n$. In FEL structure, it is assumed that the classical controller, in this case *PD* controller, is responsible for the stability of the closed-loop system in a compact set. Thus, The following conditions hold.

$$\begin{aligned} |e| &\le B_e, |\dot{e}| \le B_e, |\dot{e}| \le B_{\dot{e}}, \\ |\ddot{e}| &\le B_e, |u| \le B_u, |\dot{u}| \le B_{\dot{u}} \end{aligned} \tag{6.206}$$

It is to be mentioned that it is required that the upper bounds do exist and knowledge about their exact values are not required. In addition, the boundedness of the adaptive parameters can be guaranteed by stopping the adaptation law whenever it does not satisfy the following conditions.

$$B_{\sigma m} \le \|\sigma_1\| \le B_\sigma, B_{\sigma m} \le \|\sigma_2\| \le B_\sigma, \|\overline{m_2}\| \le B_m,$$
$$\|m_2\| \le B_m, \|\overline{m_1}\| \le B_m, \|\underline{m_1}\| \le B_m. \tag{6.207}$$

6.6.3   Sliding-Mode-Based Training Method

The *PD* controller behaves as a stabilizing controller as well as a sliding manifold. The sliding surface is defined as $S_p = \dot{e} + \chi e$, where χ is the slope of the sliding surface. The following equation shows the direct relationship between the *PD* controller and the sliding manifold.

$$S_c(u_f, u) = u_f(t) = u_n(t) + u(t) = 0 \tag{6.208}$$

which can be further modified as follows:

$$S_c(e, \dot{e}) = k_D\dot{e} + k_Pe = K_D\left(\dot{e} + \frac{k_P}{k_D}e\right) = k_DS_p, \tag{6.209}$$

where $\chi = k_p/k_D$.
The dynamic of sliding surface is achieved as

$$\dot{e} + \frac{k_P}{k_D}e = \dot{e} + \chi e = 0 \tag{6.210}$$

which can be rewritten in state space form as follows:

$$\dot{e} = -\chi e. \tag{6.211}$$

The slope χ should be selected as positive integer.

Theorem 6.7 *Let the intelligent controller consists of an IT2FLS, which works in parallel with a classical PD controller with its adaptation laws being selected as in (6.212)–(6.219), in which the conventional PD controller guarantees the stability of the system such that (6.206) holds, then $u_f(t)$, the output of the classical controller, from arbitrary initial conditions $u_f(0)$ will converge to a neighborhood near zero. Furthermore, it is possible to choose this neighborhood as small as possible. In addition, the sliding surface tends to zero as $u_f(t)$ tends to zero.*

$$\underline{\dot{m}_{1i}} = -\frac{\beta_1(\sigma_{1i})^2}{e - \frac{1}{2}\underline{m}_{1i} - \frac{1}{2}\overline{m}_{1i}} sgn(u_f), \tag{6.212}$$

$$\overline{\dot{m}_{1i}} = -\frac{\beta_1(\sigma_{1i})^2}{e - \frac{1}{2}\underline{m}_{1i} - \frac{1}{2}\overline{m}_{1i}} sgn(u_f), \tag{6.213}$$

$$\underline{\dot{m}_{2j}} = -\frac{\beta_1(\sigma_{2j})^2}{\dot{e} - \frac{1}{2}\underline{m}_{2j} - \frac{1}{2}\overline{m}_{2j}} sgn(u_f), \tag{6.214}$$

$$\overline{\dot{m}_{2j}} = -\frac{\beta_1(\sigma_{2j})^2}{\dot{e} - \frac{1}{2}\underline{m}_{2j} - \frac{1}{2}\overline{m}_{2j}} sgn(u_f), \tag{6.215}$$

$$\dot{\sigma}_{1i} = -\frac{\beta_2(\sigma_{1i})^3}{\left(e - \frac{1}{2}\underline{m}_{1i} - \frac{1}{2}\overline{m}_{1i}\right)^2} sgn(u_f), \tag{6.216}$$

$$\dot{\sigma}_{2j} = -\frac{\beta_2(\sigma_{2j})^3}{\left(\dot{e} - \frac{1}{2}\underline{m}_{2j} - \frac{1}{2}\overline{m}_{2j}\right)^2} sgn(u_f), \tag{6.217}$$

$$\dot{\theta}_{ij} = -\frac{\tilde{\zeta}_{ij}}{\sum_{i=1}^{I}\sum_{j=1}^{J}(\tilde{\zeta}_{ij})^2}\beta sgn(u_f), \tag{6.218}$$

$$\dot{\beta} = \gamma|u_f| - \gamma v\beta. \tag{6.219}$$

Proof The time derivative of ζ is achieved as

$$\dot{\underline{\zeta}}_{ij} = \frac{\left(\mu_{1i}(e)\mu_{2j}(\dot{e})\right)' \left(\sum_{i=1}^{I}\sum_{j=1}^{J}\underline{\zeta}_{ij}\right)}{\left(\sum_{i=1}^{I}\sum_{j=1}^{J}\underline{\zeta}_{ij}\right)^2}$$

$$-\frac{\left(\underline{\zeta}_{ij}\right)\left(\sum_{i=1}^{I}\sum_{j=1}^{J}\mu_{1i}(e)\mu_{2j}(\dot{e})\right)'}{\left(\sum_{i=1}^{I}\sum_{j=1}^{J}\underline{\zeta}_{ij}\right)^2},$$

$$\dot{\overline{\zeta}}_{ij} = \frac{\left(\overline{\mu_{1i}(e)\mu_{2j}(\dot{e})}\right)'\left(\sum_{i=1}^{I}\sum_{j=1}^{J}\overline{\zeta}_{ij}\right)}{\left(\sum_{i=1}^{I}\sum_{j=1}^{J}\underline{\zeta}_{ij}\right)^2}-$$

$$\frac{\left(\overline{\zeta}_{ij}\right)\left(\sum_{i=1}^{I}\sum_{j=1}^{J}\overline{\mu_{1i}(e)\mu_{2j}(\dot{e})}\right)'}{\left(\sum_{i=1}^{I}\sum_{j=1}^{J}\overline{\zeta}_{ij}\right)^2},$$

where

$$\zeta_{ij} = \exp\left(-\left(\frac{e-\frac{1}{2}m_{1i}-\frac{1}{2}\overline{m_{1i}}}{\sigma_{1i}}\right)^2 - \left(\frac{\dot{e}-\frac{1}{2}m_{2j}-\frac{1}{2}\overline{m_{2j}}}{\sigma_{2j}}\right)^2\right).$$

The normalized value of ζ_{ij} is obtained as follows:

$$\tilde{\zeta}_{ij} = \frac{\zeta_{ij}}{\sum_{i=1}^{I}\sum_{j=1}^{J}\zeta_{ij}}.$$

Thus, its time derivative is obtained as follows:

$$\dot{\zeta}_{ij} = -2\left(A_{1i}\dot{A}_{1i}-A_{2j}\dot{A}_{2j}\right)\zeta_{ij},$$

where A_{1i} and A_{2j} are defined as follows:

$$A_{1i} = \frac{(e-\frac{1}{2}\underline{m}_{1i}-\frac{1}{2}\overline{m_{1i}})}{\sigma_{1i}},$$

$$A_{2j} = \frac{(\dot{e}-\frac{1}{2}\underline{m}_{2j}-\frac{1}{2}\overline{m_{2j}})}{\sigma_{2j}},$$

$$\dot{A}_{1i} = \frac{(\dot{e}-\frac{1}{2}\dot{\underline{m}}_{1i}-\frac{1}{2}\dot{\overline{m_{1i}}})(\sigma_{1i})-(e-\frac{1}{2}\underline{m}_{1i}-\frac{1}{2}\overline{m_{1i}})(\dot{\sigma}_{1i})}{\sigma_{1i}^2},$$

$$A'_{2j} = \frac{(\ddot{e} - \frac{1}{2}\dot{m}_{2j} - \frac{1}{2}\overline{\dot{m}_{2j}})(\sigma_{2j}) - (\dot{e} - \frac{1}{2}m_{2j} - \frac{1}{2}\overline{m}_{2j})(\dot{\sigma}_{2j})}{\sigma_{2j}^2}.$$

The Lyapunov function to investigate the stability of the system is as follows:

$$V_c = \frac{1}{2}u_f^2(t) + \frac{1}{2\gamma}(\alpha - \alpha^*)^2 \quad \gamma > 0,$$

$$\dot{V}_c = u_f \dot{u}_f + \frac{1}{\gamma}\dot{\alpha}(\alpha - \alpha^*),$$

$$\dot{u}_n = \sum_{i=1}^{I}\sum_{j=1}^{J}(\dot{\theta}_{ij}\tilde{\zeta}_{ij} + \theta_{ij}\dot{\tilde{\zeta}}_{ij}) +$$
$$q\sum_{i=1}^{I}\sum_{j=1}^{J}(\dot{\theta}_{ij}\Delta\tilde{\zeta}_{ij} + \theta_{ij}\Delta\dot{\tilde{\zeta}}_{ij}) +$$
$$(1-q)\sum_{i=1}^{I}\sum_{j=1}^{J}(\dot{\theta}_{ij}\Delta\overline{\tilde{\zeta}}_{ij} + \theta_{ij}\Delta\overline{\dot{\tilde{\zeta}}}_{ij}).$$

So that

$$\dot{V}_c = u_f\left(\sum_{i=1}^{I}\sum_{j=1}^{J}(\dot{\theta}_{ij}\tilde{\zeta}_{ij}) - 2\sum_{i=1}^{I}\sum_{j=1}^{J}(A_{1i}\dot{A}_{1i} - A_{2j}\dot{A}_{2j})\theta_{ij}\tilde{\zeta}_{ij}\right.$$

$$+ 2\sum_{i=1}^{I}\sum_{j=1}^{J}\left(\theta_{ij}\tilde{\zeta}_{ij}\sum_{i=1}^{I}\sum_{j=1}^{J}(A_{1i}\dot{A}_{1i} - A_{2j}\dot{A}_{2j})\right)\tilde{\zeta}_{ij} + D + \dot{u}\right) +$$

$$\frac{1}{\gamma}\dot{\alpha}(\alpha - \alpha^*).$$

In which D is defined as follows and has a bounded value as D_m.

$$D = q\sum_{i=1}^{I}\sum_{j=1}^{J}(\dot{\theta}_{ij}\Delta\tilde{\zeta}_{ij} + \theta_{ij}\Delta\dot{\tilde{\zeta}}_{ij}) +$$
$$(1-q)\sum_{i=1}^{I}\sum_{j=1}^{J}(\dot{\theta}_{ij}\Delta\overline{\tilde{\zeta}}_{ij} + \theta_{ij}\Delta\overline{\dot{\tilde{\zeta}}}_{ij}).$$

Using some simplification to the time derivation of Lyapunov function and considering the proposed adaption law, we have

$$\dot{V}_c \leq u_f\left(\sum_{i=1}^{I}\sum_{j=1}^{J}\tilde{\zeta}_{ij}\dot{\theta}_{ij} + \dot{u}\right)$$

$$+ |u_f|\left(\frac{8B_f B_{\dot{e}}(B_e + B_c)}{B^2\sigma_m} + D_m\right) + \frac{1}{\gamma}\dot{\alpha}(\alpha - \alpha^*).$$

In which B is defined as

$$B = \left(\frac{8 B_f B_{\dot{e}} (B_e + B_c)}{B^2 \sigma_m} + D_m + B_{\ddot{u}} \right).$$

Using adaptation law for the firing strength as

$$\dot{\theta} = - \frac{\tilde{\zeta}_{ij}}{\sum_{i=1}^{I} \sum_{j=1}^{J} \tilde{\zeta}_{ij}^{2}} \beta sgn(u_f).$$

It would be concluded that

$$\dot{V}_c \leq -\alpha^* |u_f| + B|u_f| + (\alpha^* - \alpha)|u_f| + \frac{1}{\gamma} \dot{\alpha} (\alpha - \alpha^*).$$

If it is assumed that $B \leq \frac{\alpha^*}{2}$ we have

$$\dot{V}_c \leq \frac{-\alpha^*}{2} |u_f| - v \left(\alpha - \frac{1}{2} \alpha^* \right)^2 + \frac{v\alpha^* 2}{4}.$$

In order to have a negative $\dot{V}$ we must have

$$\frac{v\alpha^{*2}}{4} \leq \frac{\alpha^*}{2} |u_f| + v \left(\alpha - \frac{1}{2} \alpha^* \right)$$

which indicates that u_f has an upper bound as

$$|u_f| \leq \frac{-v\alpha^*}{2} - \frac{2v}{\alpha^*} \left(\alpha - \frac{1}{2} \alpha^* \right)^2$$

in which v is a small positive value. This mean that $u_f(t)$ will converge to a small neighborhood near zero and remains there. Futhermore, this neighborhood can be taken as small as desired and this ends the proof.

6.6.4 Implementation of the Proposed Approach on a 2-DOF Helicopter

The proposed method is programmed in an embedded microcontroller to control a 2-DOF helicopter. This real-time implementation is done using a laboratory setup namely 2-DOF helicopter model. The microcontroller that is chosen for this purpose is an ARM microcontroller, namely, LPC1768 working @100 MHz. This microcontroller is a cheap embedded solution (approximately 68$ for a development board at the year 2016).

The plant to be controlled which is a 2-DOF helicopter system is highly nonlinear, unstable dynamic system. Furthermore, there exist a high coupling between the yaw and pitch angles of this system. The four main components used for this system are as follows [6]:

1. The switching mode DC power supply (24 V, 15 A)
2. The main body of 2-DOF helicopter
3. The controller mainboard which consists of ARM microcontroller namely *LPC*1768 working @100 MHz
4. Two drivers of the DC motors.

In order to illustrate the components of this system, its rendered picture of main body is depicted in Fig. 6.5. As can be seen from the figure, this system is composed of 13 parts. Its two propellers are mounted on two DC motors. The pitch and yaw angles of the helicopter can be controlled using these two DC motors. The picture of the actual system with its drivers and the control mainboard are represented in Fig. 6.6.

The fuselage of the helicopter is 480 mm and its height is 350 mm. The diameters of the propellers are 160 mm. Two 12 V DC motors which consume 7 A at full load in 18000 rpm are used. Two 12 bit programmable magnetic encoders with part number of *AS*5045 are used to measure the pitch and yaw angles. The resolution of the sensors used in these experiments are 0.0879°.

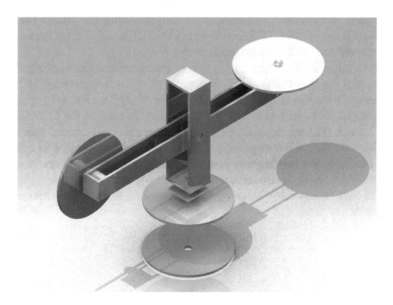

Fig. 6.5 The rendered picture of the main body of the 2-DOF helicopter

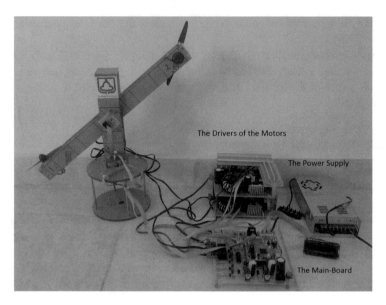

The Drivers of the Motors

The Power Supply

The Main-Board

Fig. 6.6 The 2-DOF helicopter

The heart of the mainboard which is used to implement the proposed approach is an ARM microcontroller, namely, *LPC*1768 working @100 MHz. In order to make sure that the system is isolated from the drivers during the debugging phase of the microcontroller, two relays are added to the mainboard. The feedback signals of the magnetic encoders are fed to this mainboard to control the whole system. The control signals are two PWM signals working @10 KHz, which drive the buck DC–DC converters and the direction signals of the motors.

Each driver of the DC motors used in this system are composed of a buck DC–DC converter and a H-bridge which is composed of four transistors. The DC–DC converter provides the analog DC signal to drive the DC motors with appropriate angular velocity. In order to be able to change the direction of the DC motors, H-bridge is used. The analog buck DC–DC converter is itself derived using PWM signal using the built-n PWM-generators in the ARM microcontroller. The direction signals are fed to the H-bridges to control the directions of the motors. In order to isolate the microcontroller from the power section which drive the motors, the PWM signals of the microontroller are completely isolated from the power parts using five optocoupler.

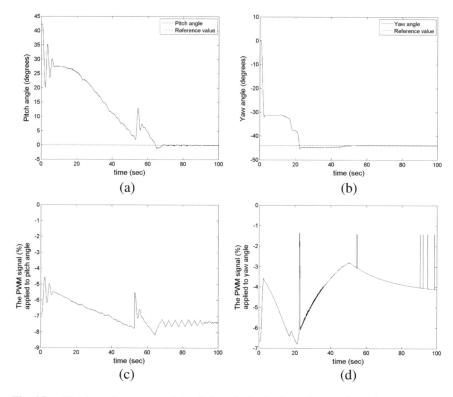

Fig. 6.7 **a** The dynamic response of the pitch angle. **b** The dynamic behavior of the yaw angle. **c** The control signal to control pitch angle. **d** The PWM signal to control yaw angle

The sample time considered for the system is set to 15 ms. The results of the implementation when the initial yaw and pitch angles are 40° and 36.5° and the reference values for yaw and pitch angles in both cases is equal to 0°, respectively, are depicted in Fig. 6.8. As can be seen from the figures, the proposed approach which is implementable in a low cost microcontroller is capable of controlling the yaw and pitch angles of the system successfully.

The response of the system in the presence of disturbance is depicted in Figs. 6.7, 6.8, 6.9, and 6.10. Moreover, in Figs. 6.8 and 6.9, weight disturbance is suddenly applied to the system at around 40th and 80th seconds. As can be seen from these figures, the proposed controller is capable of controlling the system in the presence of load disturbances. It is to be noted that as can be seen from Fig. 6.7, the control signal is not too oscillating and the oscillations in Fig. 6.9d is because of wide time range.

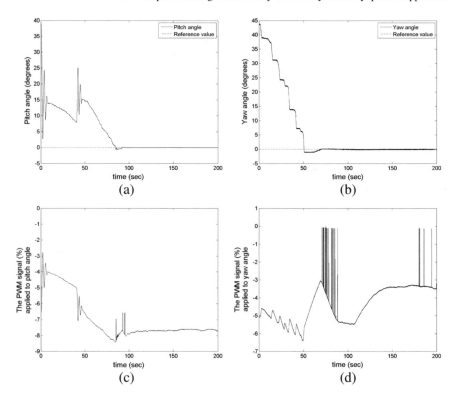

Fig. 6.8 The dynamic response of the system in the presence of disturbance at around 40 s **a** The dynamic response of the pitch angle. **b** The dynamic behavior of the yaw angle. **c** The control signal to control pitch angle. **d** The PWM signal to control yaw angle

Hence, in this case the proposed controller which is used to program an embedded microcontroller could successfully control the real-time laboratory setup. It is further shown that the proposed controller is capable of controlling the 2-DOF helicopter system even in the presence of disturbances.

Acknowledgements Authors would like to acknowledgement Alireza Jalalian Khakshour for his contribution to the implementation of the controller designed in Sect. 6.6 on a real-time system.

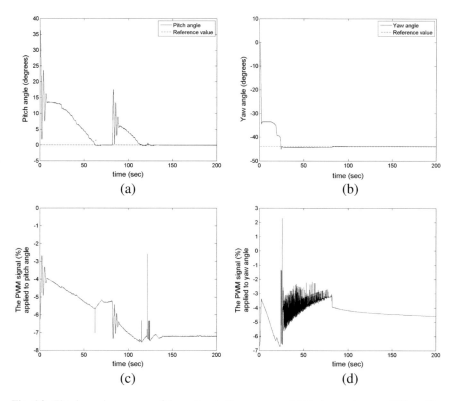

Fig. 6.9 The dynamic response of the system in the presence of disturbance at arround 80 s. **a** The dynamic response of the pitch angle. **b** The dynamic behavior of the yaw angle. **c** The control signal to control pitch angle. **d** The PWM signal to control yaw angle

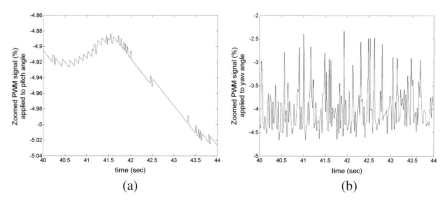

Fig. 6.10 Magnified control signals. **a** Zoomed control signal to control pitch angle. **b** Zoomed control signal to control yaw angle

References

1. Farrell, J.A., Polycarpou, M.M.: Adaptive approximation based control: unifying neural, fuzzy and traditional adaptive approximation approaches, vol. 48. Wiley, New York (2006)
2. Wang, J., Rad, A.B., Chan, P.T.: Indirect adaptive fuzzy sliding mode control: Part I: fuzzy switching. Fuzzy Sets Syst. **122**(1), 21–30 (2001)
3. Shahnazi, R., Akbarzadeh-T, M.-R.: Pi adaptive fuzzy control with large and fast disturbance rejection for a class of uncertain nonlinear systems. IEEE Trans. Fuzzy Syst. **16**(1), 187–197 (2008)
4. Chalam, V.V.: Adaptive control systems: techniques and applications. Marcel Dekker, Inc., New York (1987)
5. Kawato, M., Gomi, H.: A computational model of four regions of the cerebellum based on feedback-error learning. Biolog. Cybern. **68**(2), 95–103 (1992)
6. Khakshour, A.J., Khanesar, M.A.: Model reference fractional order control using type-2 fuzzy neural networks structure: Implementation on a 2-dof helicopter. Neurocomputing **193**, 268–279 (2016)

Chapter 7
Adaptive Network Sliding-Mode Fuzzy Logic Control Systems

7.1 Introduction

Centralized direct digital control systems have several drawbacks, for example, massive wiring requirements, difficult diagnosis, and difficult fault detection procedures. Most of these drawbacks may impose heavy costs on the maintenance of the control system. These disadvantages have given rise to the development of network control systems (NCSs). In a network-based control system, sensors, actuators, controllers, human–machine interfaces, and other possible components of the system share their data over a network [1]. An NCS connects the cyberspace to the physical one allowing remote data processing and control [2].

Since all devices on the control loop use the same network to share data, one of the advantages of NCSs is to reduce the wiring requirements. For example, the use of a network called controller area network (CAN) in BMW 850 Coupe, the wiring requirement is reduced for 2 km and the vehicle's overall weight is reduced at least to 50 kg [3]. Maintenance and fault diagnosis are also greatly eased using NCS. Moreover, the use of NCS improves the reliability of the whole process. Another prominent advantage of NCS is that it makes the extension of the system such as adding a new sensor, actuator, and data logger much simpler [4].

Most of the analysis done for a digital controller design is based on the fact that the sampling interval for a system is fixed, and the control signal is applied to the system on the basis of its sampling interval. Data collection from sensors is usually based on the same concept. Moreover, the time delay between the received data from the sensor and the applied control signal must be kept as small as possible to avoid instability in the system. Hence, real-time control applications impose some special requirements on the communication line between the controller and the plant. There exist some networks, which are specialized for real-time control purposes; examples of such are CAN and Fieldbus [5]. Furthermore, in most industrial control applications over networks, some components and/or some messages have more priority. For example, in a CAN bus, the package of data to be transmitted includes an identifier which determines the priority and importance of the message. This

© Springer Nature Switzerland AG 2021   179
M. Ahmadieh Khanesar et al., *Sliding-Mode Fuzzy Controllers*, Studies in Systems,
Decision and Control 357, https://doi.org/10.1007/978-3-030-69182-0_7

identifier guarantees that the higher priority message will be transmitted as if it is the only message existing in the network [6]. Hence, industrial networks are different from data networks and require some special considerations.

Although most industrial networks benefit from Ethernet technology, there exist fundamental differences between an industrial and a commercial network. The components used in an industrial network include various types of devices from sensors to controllers and human–machine interfaces, which are designed to monitor and control real-world processes [7]. Failure in an industrial network may be hazardous and may result in huge losses in the process and even fatalities.

From an implementation point of view, an industrial network is employed under different and hazardous conditions such as manufacturing, assembly line of automobiles, chemical refinement, electricity generation and distribution, transportation system, and water distribution, each plant having its own requirement and specific devices. The systems in the field are interconnected and may use each other's data and/or send certain commands and configurations to each other. However, the long distance between the nodes in some industrial networks, e.g., utility distribution system, makes the interconnectivity of the control network even more difficult.

7.2   Applications of Network Control Systems

7.2.1   Automotive Industry

The problem of massive wiring can be solved using a network in a typical modern automobile which connects almost all modules. Less wiring is especially important in automobile engineering as it can lessen the weight of automobiles which results in less fuel and energy consumption and less pollution. The first network protocols used in cars is CAN in which messages are given priority. This network protocol is designed such that it grants access to the network for the unit transmitting the highest priority message to ensure safe control of the vehicle. Another network which is used in modern automobiles is the vehicle area network (VAN) developed by PSA Peugeot Citroen and Renault. The most distinguished feature of VAN is that the data is encoded using Enhanced Manchester [8].

In the automotive industry, a novel technology called drive-by-wire is used, in which, the operations which were traditionally done by mechanical linkages are performed by electrical or electro-mechanical systems [9, 10]. As most of the devices in a car need the data from other nodes, the design of a drive by wire technology requires a safe communication network for the automotive devices to exchange data [11, 12]. Brake-by-wire, shift by wire, park by wire, lighting-by-wire, and door by wire can be named as other recent technologies used in modern cars [9, 10].

7.2.2 Process Control Systems

Process control systems, e.g., wastewater treatment, the dairy industries, and oil refinery, have widely used network-based control systems. In such systems, there are a huge number of sensors and actuators which cannot be handled without the use of an appropriate control network.

In a typical process network control system, there exist four levels: (1) Enterprise level, (2) Plant level, (3) Process level, and (4) Field level. Each of these levels has its own dedicated network protocols [7].

7.2.3 Fly-by-Wire

In a typical fly-by-wire system, electrical wires take the responsibility of transmitting the pilot's movements to the control surface actuators rather than mechanical linkages which traditionally used to transmit the pilot's stick movements to the actuators. The commercial aircraft $A320$, $A330/A340$ were the first commercial aircraft that used fly-by-wire technology and Boeing's first aircraft that used this technology was Boeing 777 [13]. The integration of a fly-by-wire system in a typical aircraft is done using a high-speed bidirectional data bus [14, 15].

7.2.4 Teleoperation

Teleoperation refers to the process of performing a certain task using a master and a slave system over a long distance. It has certain applications in hazardous areas and in the places, which are not easy to access [16, 17]. The long distance between the master and the slave systems makes it too difficult to connect them directly using a pair of wire or more. Even in short distances, as there usually exist more sensors and actuators in the master and the slave systems, network-based communications are more preferred [18, 19].

7.2.5 Smart Grids

Electrical load demand all over the world has experienced ultrafast growth during the past few decades. The fast-growing information and communication technology makes it possible to integrate power systems to lessen the losses in the generation of electrical energy and its transmission lines and to meet new forms of demands such as smart homes and businesses [20]. Similar to other applications in which NCSs are used, the usage of NCSs in electrical power systems makes the cost of developing these systems much lower, lessens the failure probability, and makes the installation and maintenance of these systems much easier. However, as electrical power system has a great impact on our daily life, and the security of such a network is very important. Moreover, network-induced constraints such as packet losses and time delays may lead to uncontrolled oscillations which has led to a power blackout in the past [21]. Hence, the stability analysis of NCS becomes even more important.

7.3   Common Challenges of Direct Digital Control Systems and Network Counterparts

Since network control of industrial systems is mostly done using digital signals, such systems share most of the components with direct digital controllers (DDCs) [1]. The plant is the system to be controlled whose input and output are both continuous time. However, the digital controller needs a string of digitized values to perform its process. Hence, an analog-to-digital converter is used to convert the analog signal transmitted by the transducer to digital form in every sampling time. Since the analog signal may vary during the conversion step, we need a sample and hold to guarantee that this signal is constant. The output of the analog-to-digital converter is a series of binary code, which are to be manipulated by the digital computer based on a specific algorithm. The outcome of the digital computer is then converted to a corresponding analog signal using the digital-to-analog converter. Finally, a hold system converts the signal to a continuous time, one which can be applied to the plant [22].

There are major challenges in the digital control of plants, which are common to the control of systems over the network too. These challenges include quantization, choice of sampling time and delays [22].

7.3.1   Quantization

An analog-to-digital conversion process results in a finite number of binary codes. Hence, the analog-to-digital converter rounds its analog input signal value to its nearest discrete permitted signal level. This process is called quantization. The next process is encoding in which the output of a quantized sample is described by a code. Every quantization process has a resolution. The quantization level Q is defined as the range between the two adjacent permissible levels. Let the quantization be done using n bits; the quantization level is obtained as follows [22]:

$$Q = \frac{FSR}{2^n} \tag{7.1}$$

where FSR stands for full scale range. As can be seen from Fig. 7.1, quantization causes error in the output signal. The quantization error e is defined as the difference

Fig. 7.1 The error caused by quantization

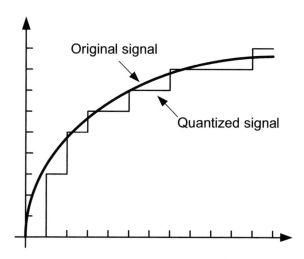

between the quantized signal and its corresponding original signal, which satisfies the following equation:

$$0 \leq |e| \leq \frac{1}{2}Q. \tag{7.2}$$

If sufficiently large number of bits are used in quantization, the quantization level becomes small resulting in a small quantization error. In this case, the quantization error can be treated as a zero-mean noise with uniform probability density function whose variance is equal to $Q^2/12$ [22].

7.3.2 Sample Time

Analog signals are digitized at certain time intervals called sample time. The selection of sample time is very important. It is highly desired to select the sample time as low as possible; however, since the processes of the collected data including analog-to-digital conversion, controller algorithm execution, and digital-to-analog conversion are needed to be done in a time interval which is much smaller than the sample time, too small sample time cannot be selected. In order to be able to reconstruct an analog signal from its discretized signal, it is required to select the sampling frequency as two times more than the highest frequency component involved in the continuous time signal. However, in practice, the stability of a closed-loop system and other design requirements may force higher sampling frequencies [22, 23].

7.3.3   Delays

The processes that are mentioned previously, such as analog-to-digital conversion, controller algorithm execution, and digital-to-analog conversion, are not done instantaneously. These processes may impose some delays on the control signal which may even disturb the stability analysis if they are neglected during stability analysis [22, 23].

7.4   Main Constraints Imposed When Controlling over Network

Since signals are needed to be digitized before they can be used in a network, the main challenges of DDC affect NCSs as well. The choice of sampling time, error caused by noise, and delay caused by processing units are shared challenges of DDC and NCS. However, there are some challenges that specifically occur in an NCS, which is the focus of this section.

7.4.1   Network-Induced Time Delays

As mentioned earlier, there are some time delays in the system which are caused by processing units. When working on a network basis, some other delays such as transmission delay over the network and wait time for queued network package to be sent can also be mentioned. However, the computation delay time is usually much less than the other two and can be neglected in a typical NCS. Delay causes phase lag in all frequency ranges and can disturb the stability analysis [1]. Hence, it must be taken into the account and appropriate action must be considered to deal with it. There are some algorithms which try to decrease the amount of data needed to be transmitted over the network to reduce network-induced time delays without changing the network structure [1, 24].

7.4.2   Packet Losses and Disorder

There exist two main transmission policies in a communication network with respect to lost data. The first policy is called TCP, in which in the case of packet loss, the data are sent again. Hence, all the packet data will be received successfully. However, it may take a long time for a packet data to be transmitted successfully. The second policy is called UDP, in which the lost data will be lost forever [25, 26].

Network traffic congestion and failure in packet transmission may cause packet losses. The possibility of packet losses is more in wireless networks [1, 27]. Packet loss makes the control system a stochastic one, which may or may not receive the current state of the system. Disconnection may also happen during control which may cause long-time packet losses. It is highly interesting for a control algorithm to work in the presence of a high possibility of packet losses.

It may even happen in a network that the network-induced time delay is much more than the sampling time which may cause disordering of the packets. The most common method with packet disordering is to discard the old packets and considering them as lost packets. However, there exist more methods in the literature that treat this problem with different controller design methods [26].

7.4.3 Variable Transmission and Sample Time

Fixed sample time, which is basically needed for most conventional direct digital control systems, cannot usually be ensured in a typical NCS. Moreover, there are multiple nodes on a single network competing to gain access. The number of nodes to transfer data on the network varies and is completely random. Hence, the time delay in a typical NCS is stochastic and varies considerably. It is required to design a robust control algorithm to deal with stability analysis problems, which may be caused by variable time delays in the network [1, 2].

7.5 Sliding-Mode Fuzzy Logic Control Techniques over Networks to Deal with Imperfection

7.5.1 Design of an Adaptive Sliding-Mode Fuzzy Logic Controller with State Prediction

Previously, an adaptive sliding-mode fuzzy logic control method for flexible spacecraft attitude control system was proposed which benefits from a predictor [28]. This controller uses a nonlinear predictor to provide time-advanced predictive states of the system. The dynamic model of the flexible spacecraft attitude control system will be composed of three subsystems, namely roll subsystem, pitch subsystem, and yaw subsystem, each of which has the following form:

$$\dot{x}_1 = x_2$$
$$\dot{x}_2 = f(x) + B(x)u + d, \tag{7.3}$$

where $\mathbf{x} \in \mathfrak{R}^2$ is the state vector of the system and $u \in \mathfrak{R}$ is the input. $f(\mathbf{x}) \in \mathfrak{R}$ and $B(\mathbf{x}) \in \mathfrak{R}$ are the unknown nonlinear functions of the system, and $d(t) \in \mathfrak{R}$ is the unknown bounded external disturbance acting on the system. It is assumed that $B(x) = B_0(x) + \Delta B(x)$ in which $B_0(x)$ is the known part and $\Delta B(x)$ is the unknown part which satisfies $|\Delta B(x)| < \Delta B_{max}$. It is assumed that $|d(t)| < d_{max}$, in which d_{max} is the upper bound of the external disturbance which is assumed to be unknown.

Three different kinds of delays are considered in this system: (1) τ_{sc}: the sensor to controller delay; (2) τ_{ca}: the controller to actuator delay; and (3) τ_c: the delay caused by the execution of the algorithm in the controller. All the delays are accumulated as the network-induced time delay shown by τ. Hence, the dynamic model of each subsystem after applying the constraints of the network is as follows:

$$\ddot{x}_1 = f(x) + B(x)u(t - \tau) + d. \tag{7.4}$$

The following variables are defined [28]:

$$\dot{x}^*(t) = f(x^*(t)) + Bu(t) \tag{7.5}$$
$$\dot{\tilde{x}}(t) = f(\tilde{x}(t)) + Bu(t) \tag{7.6}$$
$$\hat{x}(t + \hat{\tau}|t) = x^*(t) + x(t) - \tilde{x}(t) \tag{7.7}$$

where $x^*(t)$, $\tilde{x}(t)$, $x(t) \in \mathfrak{R}$ are the model state vector, the nominal state vector, and the actual plant's state vector, respectively. The parameter $\hat{x}(t + \tau|t)$ is the corrected time-advanced predictive state vector. An adaptive T1FLS $F(x^*)$ is considered, which satisfies the following equation:

$$sup_{x \in U}|F(x^*) - f(x^*)| < \varepsilon. \tag{7.8}$$

It is further supposed that $F(x^*)$ has the following form:

$$F(x^*) = \theta_f^T P_f(x^*) \tag{7.9}$$

where θ_f's are unknown adaptive parameters and $P_f(x^*)$ is a fuzzy basis function which is assumed to be known and fixed during the operation.

The sliding surface proposed for this system is constructed based on the prediction error $\hat{e}(t + \tau|t)$, given by

$$\hat{e}(t + \tau|t) = \hat{x}(t + \tau|t) - x_d(t + \tau|t) \tag{7.10}$$

and is as follows:

$$\hat{S} = K\hat{e}(t + \tau|t) \tag{7.11}$$

where $K \in \Re^{1 \times n}$ is a vector which is considered such that the sliding surface is stable.

The control signal considered to stabilize this system is as follows:

$$u = u_f + u_s \tag{7.12}$$

where

$$u_f = \frac{1}{B_0}(-K_d \hat{S} + x_d^{(n)}(t + \tau|t)$$

$$- \sum_{j=1}^{n-1} K_j \hat{e}(t + \tau|t)^{(j)} - \hat{f}(\hat{x}) \tag{7.13}$$

$$u_s = -\frac{1}{B_0}(\hat{\eta}_f + \Delta B_{max}|u| + d_{max})sat(\hat{S}) \tag{7.14}$$

and the adaptation laws for its parameters are given as follows:

$$\dot{\hat{\theta}} = \gamma_f P_f(\hat{x}(t + \tau|t))\hat{S}$$
$$\dot{\hat{\eta}}_f = \gamma_{ef}|\hat{S}|. \tag{7.15}$$

As can be seen from the adaptation laws presented in (7.15), the network-induced time delay is needed to be known. However, in the proposed approach, using Pade approximation, the network-induced time delay can be unknown and time-varying.

7.5.2 Design of an Adaptive Sliding-Mode Fuzzy Controller

The network sliding-mode control approach which uses fuzzy logic theory to model its nonlinear functions has been introduced in [29]. This controller is capable of controlling single-input–single-output (SISO) nonlinear dynamic systems in normal form in the presence of uncertain network-induced time delay. Sliding-mode controllers are widely used nonlinear controllers which benefit from the well-established mathematical proof of stability, robustness in the presence of bounded uncertainty and bounded disturbances. Since Takagi–Sugeno fuzzy logic systems (TSFLSs) are general function approximators with a high degree of flexibility, they are used to identify the nonlinear dynamic functions which exist in the structure of SISO nonlinear systems. Hence, sliding-mode fuzzy logic controllers carry the advantages of both the fuzzy logic and the sliding-mode controllers at the same time [30, 31].

The main feature of the TSFLS proposed in [29] is that it uses fuzzy logic systems to approximate the nonlinear dynamics of the system. This decreases *a priori* knowledge about the dynamics of the system considerably. Furthermore, the network-induced time delay considered in [29] is time-varying. This feature is critically important as in most networks, it is hard, if not impossible, to guarantee a constant network-induced time delay. Moreover, the use of Pade approximation makes it possible to guarantee the stability of the system using an appropriate Lyapunov function.

The nth-order nonlinear dynamic system which is considered in [29] is as follows:

$$\begin{cases} \dot{x}_1 = x_2 \\ \quad \vdots \\ \dot{x}_{n-1} = x_n \\ \dot{x}_n = f(\mathbf{x}) + g(\mathbf{x})u(t) + d(t) \end{cases} \tag{7.16}$$

where $\mathbf{x} \in \mathfrak{R}^n$ is the state vector of the system and $u \in \mathfrak{R}$ is the input. $f(\mathbf{x}) \in \mathfrak{R}$ and $g(\mathbf{x}) \in \mathfrak{R}$ are the unknown nonlinear functions of the system, and $d(t) \in \mathfrak{R}$ is the unknown bounded external disturbance acting on the system. It is assumed that $|d(t)| < D$, in which D is the upper bound of the external disturbance which is assumed to be unknown. The control of the system over the network induces an uncertain input delay as follows:

$$\begin{cases} \dot{x}_1 = x_2 \\ \quad \vdots \\ \dot{x}_{n-1} = x_n \\ \dot{x}_n = f(\mathbf{x}) + g(\mathbf{x})u(t - \tau) + d(t) \end{cases} , \tag{7.17}$$

in which τ is a time-varying parameter which represents the network-induced delay. The control objective is to make the states of the system ($\mathbf{x}$) track the desired state $\mathbf{x}_d = [x_d, \dot{x}_d, \ldots, x_d^{(n-1)}]^T$ in the presence of unknown functions, unknown disturbances, and uncertain time-varying network-induced delay. The tracking error ($\mathbf{e} \in \mathfrak{R}^n$) is defined as follows:

$$\mathbf{e} = \mathbf{x}_d - \mathbf{x} = [e, \dot{e}, \ldots, e^{(n-1)}]^T. \tag{7.18}$$

In the sliding-mode control approach, a sliding surface is designed which represents the desired behavior of the system. This variable reduces the control problem of an nth-order nonlinear system to the control of a single variable.

The sliding surface (s_p) considered in [29] is defined as follows:

$$s_p = e^{(n-1)} + \lambda_{n-2}e^{(n-2)} + \cdots + \lambda_1 e^{(1)} + \lambda_0 e, \tag{7.19}$$

in which λ's are considered such that the real parts of the roots (p) of the nth-order polynomial $P = p^{n-1} + \lambda_{n-2}p^{(n-2)} + \cdots + \lambda_1 p + \lambda_0 = 0$ are negative and hence it defines a stable trajectory for the system. It is possible to use Pade approximation to approximate the input time delay of the system, so that we have

$$\ell\{u(t-\tau)\} = exp(-\tau s)\ell\{u(t)\} = \frac{exp(-\tau s/2)}{exp(\tau s/2)}\ell\{u(t)\}$$

$$\approx \frac{(1-\tau s/2)}{(1+\tau s/2)}\ell\{u(t)\}, \tag{7.20}$$

in which $\ell\{u(t)\}$ is the Laplace transform of $u(t)$ and s is the Laplace variable. A new state variable x_{n+1} is defined as

$$\frac{1-\tau s/2}{1+\tau s/2}\ell\{u(t)\} = \ell\{x_{n+1}(t)\} - \ell\{u(t)\}. \tag{7.21}$$

Hence, the following equation is obtained for x_{n+1}:

$$\dot{x}_{n+1} = -\gamma x_{n+1} + 2\gamma u \tag{7.22}$$

where γ is defined as $\gamma = \frac{2}{\tau}$. In this way, the dynamical time-delayed system of (7.17) can be rewritten as follows [29]:

$$\begin{cases} \dot{x}_1 &= x_2 \\ &\vdots \\ \dot{x}_{n-1} &= x_n \\ \dot{x}_n &= f(\mathbf{x}) + g(\mathbf{x})(x_{n+1} - u) + d(t) \\ \dot{x}_{n+1} &= -\gamma x_{n+1} + 2\gamma u \end{cases}$$

where γ is assumed to be in the form of $\gamma = \gamma_n + \tilde{\gamma}$ in which γ_n is the nominal part of γ and is considered to be known while $\tilde{\gamma}$ is the bounded deviation of γ from its nominal value, which is caused by a variation in network-induced delay.

A rule of a *zero*-order TSFLS consists of a premise and a consequent part in the following form:

$$R^i : IF \ x_1 \ is \ A_i^1 \ and \ \dots \ and \ x_n \ is \ A_i^n \ THEN \ y = \theta_i \ \ i = 1, \dots, M, \tag{7.23}$$

in which, M is the number of rules of the TSFLS. The corresponding fuzzy MF for the fuzzy set A_i^j is $\mu_{A_i^j}$, and θ_i are crisp constant values used in the consequence part of the TSFLS. The output of the TSFLS is computed as

$$y = \frac{\sum_{i=1}^M \theta_i \prod_{j=1}^n \mu_{A_i^j}(x_j)}{\sum_{i=1}^M \prod_{j=1}^n \mu_{A_i^j}(x_j)}. \tag{7.24}$$

It is considered that

$$\xi_i(\mathbf{x}) = \prod_{j=1}^n \mu_{A_i^j}(x_j). \tag{7.25}$$

In this way, it is possible to rewrite (7.24) as

$$y = \theta^T \xi(\mathbf{x}),\qquad(7.26)$$

in which $\theta = [\theta_1, \theta_2, \ldots, \theta_M]^T$ and $\xi(\mathbf{x}) = [\xi_1, \xi_2, \ldots, \xi_M]^T$.

Two TSFLSs are considered to approximate the static nonlinear functions which exist in the structure of the dynamic system. These functions are $f(\mathbf{x})$ and $g(\mathbf{x})$. The fuzzy logic systems which approximate $f(\mathbf{x})$ and $g(\mathbf{x})$ are considered to be in $\hat{f}$ and $\hat{g}$ shown as follows:

$$\hat{f} = \theta_{\mathbf{f}}^T \xi_f(\mathbf{x}),\quad \hat{g} = \theta_{\mathbf{g}}^T \xi_g(\mathbf{x}).\qquad(7.27)$$

Furthermore, it is possible to define the optimal parameters of the fuzzy logic systems ($\theta_{\mathbf{f}}^*$ and $\theta_{\mathbf{g}}^*$) as

$$\theta_{\mathbf{f}}^* = \arg\min_{\theta_{\mathbf{f}}\in\Omega_f}[\sup_{\mathbf{x}\in\Re^n}|\hat{f}(\mathbf{x}|\theta_{\mathbf{f}}) - f(\mathbf{x})|\,]\qquad(7.28)$$

$$\theta_{\mathbf{g}}^* = \arg\min_{\theta_{\mathbf{g}}\in\Omega_g}[\sup_{\mathbf{x}\in\Re^n}|\hat{g}(\mathbf{x}|\theta_{\mathbf{g}}) - g(\mathbf{x})|\,],\qquad(7.29)$$

in which, $\hat{f}(\mathbf{x}|\theta_{\mathbf{f}})$ implies that the $\hat{f}$ is a fuzzy logic system which uses $\theta_{\mathbf{f}}$ as the parameters of the consequent part, and

$$\Omega_f = \{\theta_{\mathbf{f}}|\;\|\theta_{\mathbf{f}}\| < M_f\}\qquad(7.30)$$

$$\Omega_g = \{\theta_{\mathbf{g}}|\;\|\theta_{\mathbf{g}}\| < M_g\}\qquad(7.31)$$

in which, M_f and M_g are sufficiently large constant values. It is further assumed that $1 < \frac{\theta_{gi}^*}{\theta_{g0}^*}$, $\forall i = 1, \ldots, M$. In this context, $|\theta_{g0}^*|$ is the lower bound of $|\theta_{gi}^*|$ $\forall i = 1, \ldots, M$ and it has the same sign as θ_{gi}^*. Using these fuzzy logic approximators in (7.23), we have

$$\begin{cases} \dot{x}_1 = x_2 \\ \quad\vdots \\ \dot{x}_{n-1} = x_n \\ \dot{x}_n = \theta_{\mathbf{f}}^{*T}\xi_f(\mathbf{x}) + \theta_{\mathbf{g}}^{*T}\xi_g(\mathbf{x})(x_{n+1} - u) + w(t) + d(t) \\ \dot{x}_{n+1} = -\gamma x_{n+1} + 2\gamma u \end{cases},\qquad(7.32)$$

in which, $w(t)$ is the minimum possible approximation error and is defined as follows:

$$w(t) = f(\mathbf{x}) - \theta_{\mathbf{f}}^{*T}\xi_f(\mathbf{x}) + (g(\mathbf{x}) - \theta_{\mathbf{g}}^{*T}\xi_g(\mathbf{x}))(x_{n+1} - u).$$

As is mentioned before, it is assumed that θ_f^*, θ_g^*, x_{n+1}, $d(t)$, $w(t)$, and γ have unknown values. It is further assumed that the nominal value of γ (γ_n) is known, although its real value can deviate considerably around this parameter. We know that the disturbance term ($d(t)$) has an upper bound. Moreover, the upper bounds of norms of θ_f^* and θ_g^* namely M_f and M_g are considered to be known.

An adaptive system is used to identify the system described in (7.32) as follows:

$$
\begin{cases}
\dot{x}_1 = x_2 \\
\quad \vdots \\
\dot{x}_{n-1} = x_n \\
\dot{\hat{x}}_n = \theta_f{}^T \xi_f + \theta_g{}^T \xi_g (\hat{x}_{n+1} - u + v) \\
\dot{\hat{x}}_{n+1} = -\gamma_n \hat{x}_{n+1} + 2\gamma_n u
\end{cases}
\tag{7.33}
$$

in which (v) is an input to the identification system which is designed such that the stability of the whole system is guaranteed. This input is defined later in this section. Considering the fact that $\gamma = \gamma_n + \tilde{\gamma}$, one gets

$$
\dot{e}_{n+1} = \dot{x}_{n+1} - \dot{\hat{x}}_{n+1} = -\gamma_n e_{n+1} - \tilde{\gamma} \hat{x}_{n+1} + 2\tilde{\gamma} u
\tag{7.34}
$$

where e_{n+1} is the estimation error for x_{n+1}. Considering the sliding surface as in (7.19), the control signal and its adaptation laws are presented as the following theorem.

Theorem 7.1 ([29]) *Consider the nonlinear system to be controlled over the network in the form of (7.23). If the control signal is designed as (7.35), in which, v has the form of (7.36) and $\hat{x}_{n+1}$ is achieved from (7.37) and the parameter update rules for θ_f and θ_g are given by (7.38)–(7.43), then the sliding motion will be reached and the closed-loop system signals will be bounded and the tracking error will converge to zero asymptotically.*

$$
u(t) = (\theta_g{}^T \xi_g)^{-1} (v - x_d^{(n)} + \theta_f{}^T \xi_f + \theta_g{}^T \xi_g \hat{x}_{n+1}
\tag{7.35}
$$
$$
- \lambda_{n-2} e^{(n-1)} - \cdots - \lambda_1 e^{(2)} - \lambda_0 e^{(1)} + \hat{k}_5 sign(s_p))
$$

$$
v = -2\hat{k}_1 s_p - \frac{\hat{k}_2 s_p}{s_p^2 + \varepsilon} \hat{x}_{n+1}^2
\tag{7.36}
$$

$$
\dot{\hat{x}}_{n+1} = -\gamma_n \hat{x}_{n+1} + 2\gamma_n u
\tag{7.37}
$$

$$
\dot{\theta}_f = -\eta_2 \xi_f s_p, \quad 0 < \eta_2
\tag{7.38}
$$

$$
\dot{\theta}_{gi} =
\begin{cases}
\Upsilon_{gi} \ if \ |\theta_{g0}| < |\theta_{gi}| \ or \\
\quad if \ \theta_{gi} = \theta_{g0} \ and \ 0 < \Upsilon_{gi} sign(\theta_{gi}) , \\
0 \quad otherwise
\end{cases}
\tag{7.39}
$$

in which ξ_{gi} is the ith element of ξ_g and

$$\Upsilon_{gi} = -\eta_3 s_p \xi_{gi} \hat{x}_{n+1} + \eta_3 s_p \xi_{gi} u, \quad 0 < \eta_3 \tag{7.40}$$

$$\dot{\hat{k}}_1 = \eta_4 s_p^2, \quad 0 < \eta_4 \tag{7.41}$$

$$\dot{\hat{k}}_2 = \eta_5 \hat{x}_{n+1}^2, \quad 0 < \eta_5 \tag{7.42}$$

$$\dot{\hat{k}}_5 = \eta_6 |s_p|, \quad 0 < \eta_6. \tag{7.43}$$

It should be noted that in (7.35)–(7.43), η_2, η_3, η_4, η_5, and η_6 are the learning rates of the corresponding adaptive parameters and can be chosen by the designer to achieve the desired performance.

Proof In order to analyze the stability of the controller, the time derivative of the sliding surface (s_p) is obtained as follows:

$$\dot{s}_p = x_d^{(n)} - \theta_f^{*T} \xi_f - \theta_g^{*T} \xi_g (x_{n+1} - u) + \lambda_{n-2} e^{(n-1)} + \cdots + \lambda_1 e^{(2)} + \lambda_0 e^{(1)} - d - w. \tag{7.44}$$

The control signal (u) is considered to be as follows:

$$u = \frac{1}{\theta_g^T \xi_g} \left(v - x_d^{(n)} + \theta_f^T \xi_f + \theta_g^T \xi_g \hat{x}_{n+1} - \lambda_{n-2} e^{(n-1)} - \cdots - \lambda_1 e^{(2)} - \lambda_0 e^{(1)} - \hat{k}_5 sign(s_p) \right). \tag{7.45}$$

By applying the control signal of (7.45) to (7.44), the following equation is obtained:

$$\dot{s}_p = -\tilde{\theta}_f^T \xi_f - \theta_g^{*T} \xi_g e_{n+1} - \tilde{\theta}_g^T \xi_g \hat{x}_{n+1} + \tilde{\theta}_g^T \xi_g u + v - d - w - \hat{k}_5 sign(s_p) \tag{7.46}$$

where $\tilde{\theta}_f = \theta_f^* - \theta_f$, $\tilde{\theta}_g = \theta_g^* - \theta_g$, $\tilde{k}_1 = k_1^* - \hat{k}_1$, $\tilde{k}_2 = k_2^* - \hat{k}_2$, and $\tilde{k}_5 = k_5^* - \hat{k}_5$. It is further assumed that

$$k_1^* = \alpha_1 M_g$$

$$k_2^* = \frac{\alpha_2^{-1}}{\eta_1} \tilde{\gamma}_{max}$$

$$D + max(w) = \kappa < k_5^*, \tag{7.47}$$

in which $\|\theta_g^*\|_\infty < M_g$.

In order to prove the stability of the closed-loop system, the following candidate Lyapunov function is considered:

$$V = \frac{1}{2} s_p^2 + \frac{1}{2\eta_1} e_{n+1}^2 + \frac{1}{2\eta_2} \tilde{\theta}_f^T \tilde{\theta}_f + \frac{1}{2\eta_3} \tilde{\theta}_g^T \tilde{\theta}_g + \frac{1}{\eta_4} \tilde{k}_1^2 + \frac{1}{\eta_5} \tilde{k}_2^2 + \frac{1}{\eta_6} \tilde{k}_5^2. \tag{7.48}$$

The time derivative of the Lyapunov function is obtained as follows:

$$\dot{V} = s_p\dot{s}_p + \frac{1}{\eta_1}e_{n+1}\dot{e}_{n+1} + \frac{1}{\eta_2}\tilde{\theta}_f^T\dot{\tilde{\theta}}_f + \frac{1}{\eta_3}\tilde{\theta}_g^T\dot{\tilde{\theta}}_g + \frac{1}{\eta_4}\tilde{k}_1\dot{\tilde{k}}_1 + \frac{1}{\eta_5}\tilde{k}_2\dot{\tilde{k}}_2 + \frac{1}{\eta_6}\tilde{k}_5\dot{\tilde{k}}_5. \quad (7.49)$$

By applying (7.34) and (7.46) to (7.49), the following equation is obtained:

$$\dot{V} = s_p(-\tilde{\theta}_f^T\xi_f - \theta_g^{*T}\xi_g e_{n+1} - \tilde{\theta}_g^T\xi_g\hat{x}_{n+1} + \tilde{\theta}_g^T\xi_g u + v - w - d - \hat{k}_5 sign(s_p))$$
$$+ \frac{1}{\eta_1}e_{n+1}\dot{e}_{n+1} + \frac{1}{\eta_2}\tilde{\theta}_f^T\dot{\tilde{\theta}}_f + \frac{1}{\eta_3}\tilde{\theta}_g^T\dot{\tilde{\theta}}_g + \frac{1}{\eta_4}\tilde{k}_1\dot{\tilde{k}}_1 + \frac{1}{\eta_5}\tilde{k}_2\dot{\tilde{k}}_2 + \frac{1}{\eta_6}\tilde{k}_5\dot{\tilde{k}}_5$$
$$= \left(\frac{1}{\eta_2}\dot{\tilde{\theta}}_f^T - s_p\xi_f^T\right)\tilde{\theta}_f + \left(\frac{1}{\eta_3}\dot{\tilde{\theta}}_g^T - s_p\xi_g^T\hat{x}_{n+1} + s_p\xi_g^T u\right)\tilde{\theta}_g + s_p(-\theta_g^{*T}\xi_g e_{n+1} + v - d - w)$$
$$- \hat{k}_5|s_p| + \frac{1}{\eta_1}e_{n+1}(-\gamma_n e_{n+1} - \tilde{\gamma}\hat{x}_{n+1} + 2\tilde{\gamma}u) + \frac{1}{\eta_4}\tilde{k}_1\dot{\tilde{k}}_1 + \frac{1}{\eta_5}\tilde{k}_2\dot{\tilde{k}}_2 + \frac{1}{\eta_6}\tilde{k}_5\dot{\tilde{k}}_5. \quad (7.50)$$

The following adaptation laws for θ_f and θ_g are proposed:

$$\dot{\theta}_f = -\dot{\tilde{\theta}}_f = -\eta_2\xi_f s_p$$

$$\dot{\theta}_{gi} = -\dot{\tilde{\theta}}_{gi} = \begin{cases} \Upsilon_{gi} \ if \ |\theta_{g0}| < |\theta_{gi}| \ or \ if \theta_{gi} = \theta_{g0} \ and \ 0 < \Upsilon_{gi} sign(\theta_{gi}) \\ 0 \quad otherwise \end{cases},$$
$$\qquad (7.51)$$

where ξ_{gi} is the ith element of ξ_g and θ_{g0} has the sign equal to the sign of θ_{gi}^*, $\forall i = 1, \ldots, M$ and is the unknown lower bound of θ_g^* if $0 < \theta_{gi}^*$ or θ_{g0} is the unknown upper bound of θ_{gi}^* if $\theta_g^* < 0$ and also,

$$\Upsilon_{gi} = -\eta_3 s_p\xi_{gi}\hat{x}_{n+1} + \eta_3 s_p\xi_{gi}u. \quad (7.52)$$

The rest of the proof is done in two cases.

Case I

In the first case when $\forall i \ |\theta_{g0}| < |\theta_{gi}|$ or if $\theta_{gi} = \theta_{g0}$ and $0 < \Upsilon_{gi} sign(\theta_{gi})$, we have

$$\dot{V} = -\frac{\gamma_n}{\eta_1}e_{n+1}^2 + s_p v - s_p w - s_p d - \hat{k}_5|s_p| - \theta_g^{*T}\xi_g s_p e_{n+1}$$
$$- \frac{1}{\eta_1}e_{n+1}\tilde{\gamma}\hat{x}_{n+1} + \frac{2}{\eta_1}e_{n+1}\tilde{\gamma}u + \frac{1}{\eta_4}\tilde{k}_1\dot{\tilde{k}}_1 + \frac{1}{\eta_5}\tilde{k}_2\dot{\tilde{k}}_2 + \frac{1}{\eta_6}\tilde{k}_5\dot{\tilde{k}}_5. \quad (7.53)$$

Lemma 7.1 *For any positive value ϑ and real values for μ and λ, we have*

$$|\lambda||\mu| \le \vartheta\mu^2 + \vartheta^{-1}\lambda^2. \quad (7.54)$$

By applying Lemma 7.1, we have

$$\theta_g^{*T} \xi_g s_p e_{n+1} \leq \alpha_1 M_g s_p^2 + \alpha_1^{-1} M_g e_{n+1}^2, \qquad 0 < \alpha_1 \tag{7.55}$$

so that

$$\dot{V} \leq -\frac{\gamma_n}{\eta_1} e_{n+1}^2 + s_p v + \kappa |s_p| - \hat{k}_5 |s_p| + \alpha_1 M_g s_p^2 + \alpha_1^{-1} M_g e_{n+1}^2 + \frac{\alpha_2 \tilde{\gamma}_{max}}{\eta_1} e_{n+1}^2$$
$$+ \frac{\alpha_2^{-1} \tilde{\gamma}_{max}}{\eta_1} \hat{x}_{n+1}^2 + \frac{2 \tilde{\gamma}_{max} \alpha_3}{\eta_1} e_{n+1}^2 + \frac{2 \alpha_3^{-1} \tilde{\gamma}_{max}}{\eta_1} u^2 + \frac{1}{\eta_4} \tilde{k}_1 \dot{k}_1 + \frac{1}{\eta_5} \tilde{k}_2 \dot{k}_2 + \frac{1}{\eta_6} \tilde{k}_5 \dot{k}_5, \tag{7.56}$$

in which $\tilde{\gamma}_{max}$ is the maximum allowable deviation of γ from its nominal value γ_n, and $\kappa = D + max(w)$ so that

$$\dot{V} \leq -\frac{\gamma_n}{\eta_1} e_{n+1}^2 + s_p v + k_5^* |s_p| - \hat{k}_5 |s_p| + k_1^* s_p^2 + k_3^* u^2$$
$$+ k_2^* \hat{x}_{n+1}^2 + k_4^* e_{n+1}^2 + \frac{1}{\eta_4} \tilde{k}_1 \dot{k}_1 + \frac{1}{\eta_5} \tilde{k}_2 \dot{k}_2 + \frac{1}{\eta_6} \tilde{k}_5 \dot{k}_5, \tag{7.57}$$

where k_1^*, k_2^*, and k_5^* are as in (7.47) and

$$k_3^* = \frac{2 \alpha_3^{-1}}{\eta_1} \tilde{\gamma}_{max}$$
$$k_4^* = \alpha_1^{-1} M_g + \frac{\alpha_2}{\eta_1} \tilde{\gamma}_{max} + \frac{2 \alpha_3}{\eta_1} \tilde{\gamma}_{max}. \tag{7.58}$$

Let the robust term in the control signal $(v(t))$ be $(v = -2\hat{k}_1 s_p - \frac{\hat{k}_2 s_p}{s_p^2 + \varepsilon} \hat{x}_{n+1}^2)$ with ε being a small positive value; we have

$$\dot{V} \leq -\frac{\gamma_n}{\eta_1} e_{n+1}^2 - \hat{k}_1 s_p^2 + \tilde{k}_1 s_p^2 + \tilde{k}_2 \hat{x}_{n+1}^2 + \hat{k}_2 \frac{\varepsilon \hat{x}_{n+1}^2}{s_p^2 + \varepsilon} + \tilde{k}_5 |s_p|$$
$$+ k_3^* u^2 + k_4^* e_{n+1}^2 + \frac{1}{\eta_4} \tilde{k}_1 \dot{k}_1 + \frac{1}{\eta_5} \tilde{k}_2 \dot{k}_2 + \frac{1}{\eta_6} \tilde{k}_5 \dot{k}_5. \tag{7.59}$$

Considering the adaptation laws as

$$\dot{\hat{k}}_1 = -\dot{\tilde{k}}_1 = -\eta_4 s_p^2$$
$$\dot{\hat{k}}_2 = -\dot{\tilde{k}}_2 = -\eta_5 \hat{x}_{n+1}^2$$
$$\dot{\hat{k}}_5 = -\dot{\tilde{k}}_5 = -\eta_6 |s_p|, \tag{7.60}$$

one gets

$$\dot{V} \leq -\frac{\gamma_n}{\eta_1} e_{n+1}^2 - \hat{k}_1 s_p^2 + k_4^* e_{n+1}^2 + \hat{k}_2 \frac{\varepsilon \hat{x}_{n+1}^2}{s_p^2 + \varepsilon} + k_3^* u^2. \tag{7.61}$$

It is possible to take the design value η_1 as

$$\eta_1 \leq \frac{\gamma_n}{2k_4^*}. \tag{7.62}$$

Furthermore, considering (7.58) and the fact that $\gamma_n = \frac{2}{\tau}$, we have

$$\alpha_1^{-1} M_g \eta_1 + \alpha_2 \tilde{\gamma}_{max} + 2\alpha_3 \tilde{\gamma}_{max} < \frac{1}{\tau}. \tag{7.63}$$

It should be noted that if τ, the nominal value of the network-induced delay, is too large, $\frac{1}{\tau}$ becomes too small; it is possible that there exists no selection for the parameter η_1 and hence, the stability conditions may not be satisfied. This equation also introduces an upper bound to the maximum permissible value for the network-induced delay in the system as follows:

$$\tau < \frac{1}{\alpha_2 \tilde{\gamma}_{max} + 2\alpha_3 \tilde{\gamma}_{max} + \alpha_1^{-1} M_g \eta_1} \tag{7.64}$$

and the time derivative of the Lyapunov function is finally achieved as

$$\dot{V} \leq -\frac{\gamma_n}{2\eta_1} e_{n+1}^2 - \hat{k}_1 s_p^2 + \hat{k}_2 \frac{\varepsilon \hat{x}_{n+1}^2}{s_p^2 + \varepsilon} + k_3^* u^2. \tag{7.65}$$

Case II

In the second case, when for one of the θ_{gi}'s we have $\theta_{g0} = \theta_{gi}$ and $\Upsilon_{gi} sign(\theta_{gi}) < 0$. Then, using the similar calculations as in the first case, one gets

$$\dot{V} \leq -\frac{\gamma_n}{2\eta_1} e_{n+1}^2 - \hat{k}_1 s_p^2 + \hat{k}_2 \frac{\varepsilon \hat{x}_{n+1}^2}{s_p^2 + \varepsilon} + k_3^* u^2 + (\theta_{gi}^* - \theta_{g0}) \Upsilon_{gi}. \tag{7.66}$$

In this case, $(\theta_{gi}^* - \theta_{g0}) \Upsilon_{gi}$ is negative and hence, the addition of this term makes the time derivative of the Lyapunov function more negative. If more θ_{gi}'s satisfy the second condition, using similar analysis, a negative term is added to the time derivative of the Lyapunov function and we have

$$\dot{V} \leq -\frac{\gamma_n}{2\eta_1} e_{n+1}^2 - \hat{k}_1 s_p^2 + \hat{k}_2 \frac{\varepsilon \hat{x}_{n+1}^2}{s_p^2 + \varepsilon} + k_3^* u^2, \tag{7.67}$$

which means that s_p and e_{n+1} converge asymptotically until s_p and e_{n+1} belong to a neighborhood of zero, in which,

Fig. 7.2 The aero-pendulum
system

$$\frac{\gamma_n}{2\eta_1}e_{n+1}^2 + \hat{k}_1 s_p^2 < \hat{k}_2 \frac{\varepsilon \hat{x}_{n+1}^2}{s_p^2 + \varepsilon} + k_3^* u^2. \qquad (7.68)$$

This neighborhood is called S. It should also be mentioned that ε is a design parameter and can be chosen as small as desired. Furthermore, it is possible to choose a
sufficiently large value for α_3 so that k_3^* becomes as small as desired. In this way, it
is possible to compensate the effect of deviation of network-induced delay from it
nominal value. This means that it is possible to make the neighborhood S very small.
However, since α_3 is also directly affecting k_4^* and a large value for it makes k_4^* too
large, it is not possible to use too large values for α_3. This introduces an upper bound
to the maximum deviation of network-induced delay from its nominal value.

7.5.3 Simulation Results and Discussions

In order to investigate the applicability of the proposed controller in controlling
benchmark problems, the network sliding-mode fuzzy logic control system designed
in the previous section is simulated for the tracking and balancing problems of an
aero-pendulum (see Fig. 7.2). The dynamic equations of motion governing the system
are as follows [32]:

$$\begin{aligned}
\dot{x}_1 &= x_2 \\
\dot{x}_2 &= -\frac{g}{L}sin(x_1) - \frac{c}{ML^2}x_2 + \frac{K}{ML}u
\end{aligned} \qquad (7.69)$$

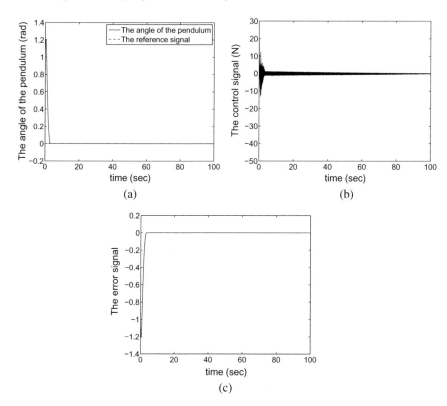

Fig. 7.3 The regulation response of the inverted pendulum subject to disturbance and time-varying network-induced delay. **a**: The angle of the inverted pendulum $x_1(t)$. **b**: The control signal $u(t)$. **c**: The error signal $e(t)$

where x_1 stands for the angle of the pendulum, zero being its value when the pendulum is straight down. The angular velocity is denoted by x_2, $g = 9.8\,\mathrm{m/s}^2$ is the gravity constant, $M = 1\,\mathrm{kg}$ is the mass of the pendulum, L is the length of the pendulum, and u is the lift force generated by the propeller. The inputs to both FLSs is considered to be x_1 with its MFs is selected as follows:

$$\mu_1(x_1) = exp\left(-\frac{(x_1 + 1)^2}{2 \times 0.33^2}\right)$$

$$\mu_2(x_1) = exp\left(-\frac{x_1^2}{2 \times 0.33^2}\right)$$

$$\mu_3(x_1) = exp\left(-\frac{(x_1 - 1)^2}{2 \times 0.33^2}\right). \tag{7.70}$$

The sample time of the system is considered to be 0.001 s and the nominal value of network-induced delay in the system (τ_n) is considered to be 0.0002 s. The initial

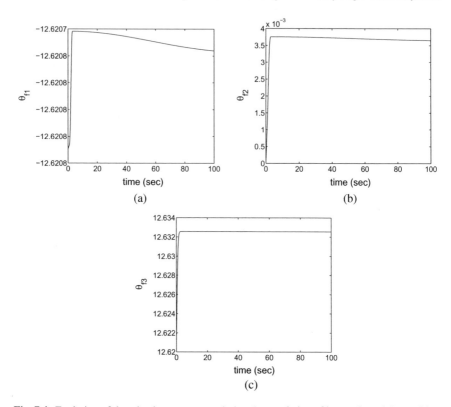

Fig. 7.4 Evolution of the adaptive parameters during the regulation of inverted pendulum subject to disturbance and time-varying network-induced delay. **a**: Evolution of $\hat{x}_{n+1}(t)$. **b**: Evolution of $\|\theta_{\mathbf{f}}\|$. **c**: Evolution of $\|\theta_{\mathbf{g}}\|$

conditions for the parameters of the consequent part parameters of the FLSs are considered as follows:

$$\theta_{g1}(0) = 0.75, \quad \theta_{g2}(0) = 1.46, \quad \theta_{g3}(0) = 0.75$$
$$\theta_{f1}(0) = -12.62, \quad \theta_{f2}(0) = 0, \quad \theta_{f3}(0) = 12.62. \tag{7.71}$$

7.5.4 Time-Varying Network-Induced Time Delay Case

As mentioned earlier, the controller is capable of controlling the system in the presence of uncertain network-induced time delays. In order to investigate this feature, it is assumed that its true value uniformly varies in an interval as follows $(0.0001 < \tau < 0.0003)$. The regulation response of the inverted pendulum is shown

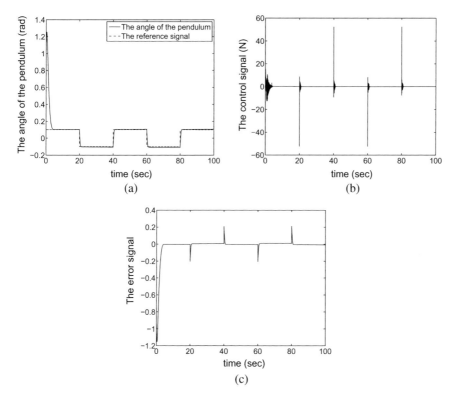

Fig. 7.5 Tracking response of the inverted pendulum subject to disturbance and time-varying network-induced delay. **a**: Angle of the inverted pendulum $x_1(t)$. **b**: Control signal $u(t)$. **c**: Error signal $e(t)$

in Fig. 7.3. The initial conditions considered for this system are selected to be equal to $x_1 = 1$, $x_2 = 0$, and the disturbance is considered to be $(d(t) = 0.1sin(t/40))$.

It is further assumed that the system is subject to wind. As can be seen from this figure, the network control system can regulate the aero-pendulum satisfactorily. Furthermore, the evolution of adaptive parameters during the regulation of the inverted pendulum is shown in Fig. 7.4. As can be seen from the figures, the parameters of the FLSs adaptively change and finally, converge to an appropriate value.

It is to be noted that since the control signals considered in this paper are compensating the upper bounds of existing uncertainties, in order to obtain satisfactory results, the fine-tuning of learning rates may be required. Furthermore, $\hat{x}_{n+1}$ is an important parameter and oscillations in this parameter may cause too oscillatory control signal; in order to avoid these oscillations, the bandwidth of u in (7.37) is limited. The tracking performances of the system when the reference signals are square and sinusoidal waves are presented in Figs. 7.5 and 7.6, respectively. The initial conditions to test the tracking performance are considered as $x_1 = 1$, $x_2 = 0$, and the disturbance is considered to be $(d(t) = 0.1sin(t/40))$. The rest of the conditions

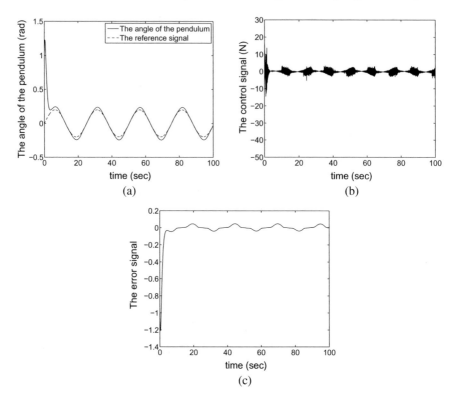

Fig. 7.6 Tracking response of the inverted pendulum when the reference signal is a sinusoidal wave and the system subject to disturbance and time-varying network-induced delay. **a**: Angle of the inverted pendulum $x_1(t)$. **b**: Control signal $u(t)$. **c**: Error signal $e(t)$

of the system under control are kept the same as in the regulation part. The performances of the systems are satisfactory, especially when tracking the square wave. As can be seen in the figure, sometimes large control signals are needed to guarantee the robustness of the control signal in the presence of time-varying network-induced delays.

It is further observed in the simulations that if the nominal value of the network-induced time delay (τ) in the system becomes too large, instability may occur. These results support the logical expectation of the existence of an upper bound for the parameter τ in the system. The existence of this upper bound can be shown explicitly using (7.64).

The simulation results on an aero-pendulum illustrate that the usage of the sliding-mode fuzzy controller makes it possible to benefit from the well-established mathematical techniques to prove the stability of the system. The stability of adaptation laws and asymptotic convergence of tracking error to a neighborhood near zero is analyzed using an appropriate Lyapunov function. The proposed control method is robust to time-varying delays of the network and bounded input disturbances.

References

1. Zhang, L., Gao, H., Kaynak, O.: Network-induced constraints in networked control systems-a survey. IEEE Trans. Ind. Inf. **9**(1), 403–416 (2013)
2. Gupta, R.A., Chow, M.-Y.: Overview of networked control systems. In: Networked Control Systems, pp. 1–23. Springer, Berlin (2008)
3. Direct, S.: A brief explanation of can bus. Technical Report (2015)
4. Pfeiffer, O., Ayre, A., Keydel, C.: Embedded Networking with CAN and CANopen. Copperhill Media (2008)
5. Knapp, E.D., Langill, J.T.: Industrial Network Security: Securing critical infrastructure networks for smart grid, SCADA, and other Industrial Control Systems. Syngress (2014)
6. Buchanan, B.: The Handbook of Data Communications and Networks: vol. 2. Springer Science & Business Media, Berlin (2010)
7. Stouffer, K., Falco, J., Scarfone, K.: Guide to industrial control systems (ics) security. NIST Special Publ. **800**(82), 16 (2011)
8. Zurawski, R.: Embedded Systems Handbook. CRC Press, Industrial Information Technology (2005)
9. Kelling, N.A., Leteinturier, P.: X-by-wire: opportunities, challenges and trends. Technical Report, SAE Technical paper (2003)
10. Randy Frank Contributing. X-by-wire: For power, x marks the spot (2004)
11. Pan, Y.-J., Canudas-de Wit, C., Sename, O.: A new predictive approach for bilateral teleoperation with applications to drive-by-wire systems. IEEE Trans. Robot. **22**(6), 1146–1162 (2006)
12. Bertoluzzo, M., Buja, G., Zuccollo, A.: Design of drive-by-wire communication network for an industrial vehicle. In: *2004 2nd IEEE International Conference on Industrial Informatics, 2004. INDIN'04*, pp. 155–160. IEEE (2004)
13. Spitzer, C.R., Spitzer, C.: Boeing b-777: Fly-by-wire flight controls. In: Digital Avionics Handbook, pp. 203–216. CRC Press, New York (2000)
14. Robert Bosch GmBH. Can specification version 2.0. (1991)
15. Tomczyk, A.: Experimental fly-by-wire control system for general aviation aircraft. Future **7**, 9 (2003)
16. Lichiardopol, S.: A survey on teleoperation. Department of Mechanical Engineering, Dynamics Control Group, Technische Universiteit Eindhoven, Eindhoven, Department of Mechanical Engineering, Dynamics Control Group, The Netherlands, Technical Report DCT2007, 155 (2007)
17. Gupta, G.S., Mukhopadhyay, S.C., Messom, C.H., Demidenko, S.N.: Master-slave control of a teleoperated anthropomorphic robotic arm with gripping force sensing. IEEE Trans. Instrum. Meas. **55**(6), 2136–2145 (2006)
18. Fraisse, P., Leleve, A.: Teleoperation over ip network: network delay regulation and adaptive control. Auton. Robots **15**(3), 225–235 (2003)
19. Chopra, N., Berestesky, P., Spong, M.W.: Bilateral teleoperation over unreliable communication networks. IEEE Trans. Control Syst. Technol. **16**(2), 304–313 (2008)
20. Singh, A.K., Singh, R., Pal, B.C.: Stability analysis of networked control in smart grids. IEEE Trans. Smart Grid **6**(1), 381–390 (2015)
21. Andersson, G., Donalek, P., Farmer, R., Hatziargyriou, N., Kamwa, I., Kundur, P., Martins, N., Paserba, J., Pourbeik, P., Sanchez-Gasca, J., et al.: Causes of the 2003 major grid blackouts in north america and europe, and recommended means to improve system dynamic performance. IEEE Trans. Power Syst. **20**(4), 1922–1928 (2005)
22. Ogata, K.: Discrete-Time Control Systems, vol. 2. Prentice Hall, Englewood Cliffs (1995)
23. Benjamin, C.K.: Digital Control Systems. Oxford University Press Inc, Oxford (1995)
24. Hespanha, J.P., Naghshtabrizi, P., Xu, Y.: A survey of recent results in networked control systems. Proc. IEEE **95**(1), 138–162 (2007)
25. Stallings, W.: Data and Computer Communications. Pearson/Prentice Hall, Upper Saddle River (2007)

26. Zhao, Y.-B., Sun, X.-M., Zhang, J., Shi, P.: Networked control systems: the communication basics and control methodologies. Math. Problems Eng. (2015)
27. Ulusoy, A., Gurbuz, O., Onat, A.: Wireless model-based predictive networked control system over cooperative wireless network. IEEE Trans. Ind. Inf. **7**(1), 41–51 (2011)
28. Dong, C., Lijie, X., Chen, Y., Wang, Q.: Networked flexible spacecraft attitude maneuver based on adaptive fuzzy sliding mode control. Acta Astronaut. **65**(11), 1561–1570 (2009)
29. Khanesar, M.A., Kaynak, O., Yin, S., Gao, H.: Adaptive indirect fuzzy sliding mode controller for networked control systems subject to time-varying network-induced time delay. IEEE Trans. Fuzzy Syst. **23**(1), 205–214 (2015)
30. Kaynak, O., Erbatur, K., Ertugnrl, M.: The fusion of computationally intelligent methodologies and sliding-mode control-a survey. IEEE Trans. Ind. Electron. **48**(1), 4–17 (2001)
31. Xinghuo, Y., Kaynak, O.: Sliding-mode control with soft computing: a survey. IEEE Trans. Ind. Electron. **56**(9), 3275–3285 (2009)
32. Yu, P.A., et al.: Introduction to Control of Oscillations and Chaos, vol. 35. World Scientific, Singapore (1998)

Chapter 8
Sliding-Mode Fuzzy Logic Teleoperation Controllers

Control of a system—for example, operation of a robot—through a communication link from a distant location is called teleoperation. Remote operations in hazardous and unreachable areas are crucial and inevitable; examples of such areas are complex and challenging tasks in disaster areas, space robotic applications, remotely operated vehicles (e.g., deep ocean robotics), and remote surgery applications. In different applications, the distance can vary from a few centimeters (operating fine) to millions of kilometers (used in space exploration).

The components of a typical teleoperation system are a master robot, a communication line, a slave robot, an external task environment, and a human operator, which controls the whole system.

Teleoperation can be categorized as "unilateral" and "bilateral". If only the master motion and/or force is transmitted to the slave site, the teleoperation system is said to be unilateral. On the other hand, if the motion and/or force information are transmitted between slave and master sites, the teleoperation system is called bilateral. With the rapid growth of the Internet and high increase in its reliability, as the reliable communication between the master and the slave systems, teleoperation technology has found its place in many applications, ranging from surgery to aerospace applications [1, 2].

In bilateral teleoperation, not only does the system control a remote manipulator but also the forces exerted on the robot in a remote environment are transmitted to the master site. Other than guarantee in the stability of the system, it is highly desired to keep the response of slaves as close as possible to the master system. Such feature is called "transparency", which is a prominent objective that should be satisfied throughout teleoperation [3, 5]. In other words, the teleoperation system is said to be completely transparent if the master and slave states are equal, and the force met by the master and slave systems are exactly the same. To address stability and transparency problems, many methods have already been investigated in the literature.

© Springer Nature Switzerland AG 2021   203
M. Ahmadieh Khanesar et al., *Sliding-Mode Fuzzy Controllers*, Studies in Systems,
Decision and Control 357, https://doi.org/10.1007/978-3-030-69182-0_8

Since the distance between the master and slave systems may be hundreds of kilometers, in most applications, a dedicated line between the master and the slave systems does not exist. Time delay is frequently observed in such teleoperation systems as a result of a network connection between them which is time varying and uncertain. Network-induced time delay results in a more difficult control problem. Bilateral teleoperation systems in the presence of variable time delay have also been considered in previous works [3]; and the stability analysis of the system is investigated.

In this chapter, sliding-mode fuzzy controllers are used to obtain transparency between master and slave systems. The fuzzy logic system considered in this case is a model-free rule-based approach that is capable of dealing with a large class of nonlinear systems. It is shown that the proposed approach is capable of obtaining transparency between master and slave systems with high performance. The use of fuzzy logic systems makes it possible to control the system without exact knowledge of the nonlinear dynamics of the system.

8.1 Sliding-Mode Fuzzy Logic Teleoperation of Robotic Manipulator

8.1.1 The Dynamics of Teleoperation Systems

The nolinear dynamic equations of a teleoperation system, which contains two n -degrees of freedom robots, are as follows:

$$M_m(q_m)\ddot{q}_m + C_m(q_m, \dot{q}_m) + g_m(q_m) = F + \tau_m \tag{8.1}$$

$$M_s(q_s)\ddot{q}_s + C_s(q_s, \dot{q}_s) + g_s(q_s) = F - \tau_s \tag{8.2}$$

where $i = m, s$ representing master and slave robots, respectively; $q_i, \dot{q}_i, \ddot{q}_i \in R^n$ denote the joint positions, velocities, and accelerations, respectively; $M_i(q_i) \in R^{n \times n}$ represents the Coriolis and centrifugal effects, $g_i(q_i) \in R^n$ the gravitational torques, and $\tau \in R^n$ the control input torques. Finally, $F_h, F_e \in R^n$ indicate the forces at the joints due to the forces exerted by the human operator and the task environment, respectively.

In this example problem, we consider two 2DOF planar robots [4] as shown in Fig. 8.1, as the master and slave systems. As the planar robots act on the xy-plane, there exist no gravitational terms in the equations of motion. In this way, (8.1) and (8.2) are reduced to

$$M_m(q_m)\ddot{q}_m + C_m(q_m, \dot{q}_m) = F_h + \tau_m \tag{8.3}$$

$$M_m(q_m)\ddot{q}_m + C_s(q_s, \dot{q}_s) = F_e - \tau_s. \tag{8.4}$$

Fig. 8.1 A 2DOF robot

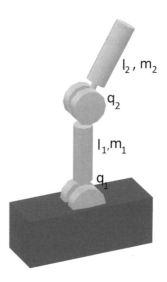

8.1.2 Controller Design

A fuzzy logic rule-based system is designed to teleoperate a 2DOF robotic arm in a way that the increment in the control signal is directed toward making $s\dot{s}$ negative. Such a controller can result in an stable closed-loop system. It is assumed that the slave system may or may not have contact with objects in a real environment. It is called a contact motion when the slave system has an interaction with the environment and is called free motion otherwise.

The main goals of the designed teleoperation controller are as follows [5]:

1. To stabilize the closed-loop system in free and contact motions, in the presence of time delays induced by the teleoperation system.

2. Accurate transparent position tracking for the master and slave robots in the presence of time delays resulting in human force to converge to environmental force. If complete transparency is obtained in contact motion, the operator can sense the reflecting force of the environment. Such transparency may also be scaled to avoid excessive force on a human operator. For instance, if the operator is manipulating a heavy object, a force sensed by the slave needs to be scaled by a factor smaller than one. On the other hand, in some precise applications, e.g., in an eye surgery, the scaling factor of force sensed by a slave needs to be scaled by a factor larger than one as an excessive force in the slave part may damage the patient's eye.

The position synchronization errors between the master and the slave systems are defined as follows:

$$e_m = q_s(t - T) - q_m(t) \tag{8.5}$$
$$e_s = q_m(t - T) - q_s(t). \tag{8.6}$$

In order to obtain full synchronization between the master and the slave systems, the following equations must be satisfied:

$$\lim_{t \to \infty} e_m(t) = \lim_{t \to \infty} e_s(t) = 0. \tag{8.7}$$

In addition, the force tracking errors in contact motion are as follows:

$$e_{fm} = F_e(t - T) - F_h(t) \tag{8.8}$$
$$e_{fs} = F_h(t - T) - F_e(t). \tag{8.9}$$

In order to guarantee the complete transparency, the following equations must be satisfied:

$$\lim_{t \to \infty} e_{fm}(t) = \lim_{t \to \infty} e_{fs}(t) = 0. \tag{8.10}$$

After $\ddot{q}_m = \ddot{q}_s = \dot{q}_m = \dot{q}_s = 0$.

8.1.2.1   Sliding-Mode Controller

Let the single input nonlinear system be defined as follows:

$$
\begin{aligned}
\dot{x}_1 &= x_2 \\
\dot{x}_2 &= x_3 \\
&\;\;\vdots \\
\dot{x}_n &= f(x) + b(x)u + d(t) \\
y(t) &= x_1(t) \qquad\qquad\qquad for\ t \geq 0
\end{aligned}
\tag{8.11}
$$

where $x^T = [x_1, x_2, \ldots, x_n]$ is the state vector, $u(t)$ is the manipulated input, x is the output state, and $d(t)$ represents the external disturbances. If $y_d(t)$ is the desired output, then (8.11) can be transformed into the following new state equations with the definition of the error signal as $e_1 = y_d - y$ and its time derivatives as the state variables:

$$
\begin{aligned}
\dot{e}_1 &= e_2 \\
\dot{e}_2 &= e_3 \\
&\;\;\vdots \\
\dot{e}_n &= f(e) + b(e)u + d(t) \\
y(t) &= e_1(t) \qquad\qquad\qquad for\ t \geq 0.
\end{aligned}
\tag{8.12}
$$

With the new definition for the state vector as e, the problem is converted to a tracking problem, which means that the convergence of the new states to *zero* forces our state vector (x) to reach the desired states y_d. A sliding manifold can be represented as follows:

$$s = c_1 e_1 + c_2 e_2 + \cdots + e_n = 0. \tag{8.13}$$

Notice that (8.13) is equivalent with the following definition of sliding hyperplane from [6]:

$$s(x) = \left(\frac{d}{dt} + \lambda\right)^{n-1} e. \tag{8.14}$$

For the motion of the system to be stable, the parameter λ must be strictly positive constant. Hence, in order to remotely control the system successfully, it is required that the first-order variable s be stabilized to *zero*. After reaching mode is passed and states of the system converge to the sliding manifold, the dynamic behaviors of the system are represented by the following equation.

$$\dot{e}_i = e_{i+1}, \quad i = 1, 2, \ldots, n - 1$$
$$\dot{e}_{n-1} = -c_1 e_1 - c_2 e_2 - \cdots - c_{n-1} e_{n-1}. \tag{8.15}$$

If the Lyapunov function for the system is considered to be $V = \frac{1}{2}s^2$, in order to maintain the stability of the system, it is required to design the control signal such that $s\dot{s} < 0$. Hence, it is required that when s is positive, then its time derivative must be negative to maintain the stable behavior of the system.

The fuzzy logic system acts on the sliding line and its time derivative, s and $\dot{s}$, and its output is the variation in the control signal, Δu. The support set of the fuzzy logic system for its inputs and output are considered as the interval of $[-1, \ 1]$. However, it is possible to use the scaling factor for those cases when inputs fall outside this region to push them within this region to be able to find an appropriate control signal. Such scaling in the controller output may also be required to speed up the response of the system and increase the steady state accuracy of the system. The seven membership functions (MFs) used for the fuzzy system are Gaussian MFs, which are depicted in Fig. 8.2. The corresponding labeling for MFs is presented in Table 8.1. The rules of the fuzzy logic system, which are designed to make $s\dot{s}$ negative, are presented in Table 8.2.

In order to explain the basic idea behind these kinds of rules for the fuzzy system, let us consider the time derivative of the sliding manifold as follows.

$$\dot{s} = \sum_{i=1}^{n-1} c_i e_{i+1} - f(e) - b(e)u - d(t). \tag{8.16}$$

If both sides of this equation are multiplied by s, the following equation is obtained:

$$s\dot{s} = \sum_{i=1}^{n-1} c_i e_{i+1} s - f(e)s - b(e)us - d(t)s. \tag{8.17}$$

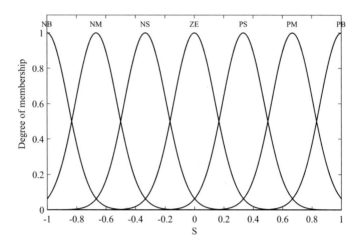

Fig. 8.2 Inputs/Output membership functions

Table 8.1 Literature of membership functions

Fuzzy variables	Description
PB	Positive Big
PM	Positive Medium
PS	Positive Small
ZE	Zero
NS	Negative Small
NM	Negative Medium
NB	Negative Big

Table 8.2 Rules of sliding-mode fuzzy controller

		S						
		PB	PM	PS	ZE	NS	NM	NB
S	PB	PB	PB	PB	PB	PM	PS	ZE
	PM	PB	PB	PB	PM	PS	ZE	NS
	PS	PB	PB	PM	PS	ZE	NS	NM
	ZE	PB	PM	PS	ZE	NS	NM	NB
	NS	PM	PS	ZE	NS	NM	NB	NB
	NM	PS	ZE	NS	NM	NB	NB	NB
	NB	ZE	NS	NM	NB	NB	NB	NB

Here, it is assumed that $b(e) > 0$ for any e, as it is mentioned earlier that the rules given in Table 8.2 are designed such that $s\dot{s} < 0$. As can be seen from (8.17), if $s > 0$, increase in the value of u leads to decrease in $s\dot{s}$. On the other hand, if $s < 0$, decrease in u results in a decrease in $s\dot{s}$. This is the main rule that is used to construct the fuzzy logic system upon. Some of the rules in the fuzzy system are described here to see how these rules are written.

The first rule is as follows:

IF S IS NS AND $\dot{S}$ is NB then Δu is NB

In this case, since the signs of s and $\dot{s}$ are the same, the stability conditions are not satisfied. In order to make $s\dot{s}$ negative, the parameter $\dot{s}$ must be positive. Hence, the change in the control signal must be negative, and it must have a large value to change the sign of $\dot{s}$ and guarantee the stability conditions.

Consider another rule as follows:

IF s IS PM AND $\dot{s}$ is NB then Δu is NS

In this case, $S\dot{S}$ is already a negative large value, and needs to be decreased. Hence, the control signal must be negative. However, since the signs of s and $\dot{s}$ are different, the change in control signal must be small to avoid chattering.

The last rule is as follows:

IF S IS PB AND $\dot{S}$ is PB then Δu is PB

In this case, $s\dot{s}$ has a very large positive value. In order to make it negative, it is required to take a large positive change in the control signal.

8.1.3 Simulation Results

In this section, the fuzzy logic-based bilateral control strategy is presented on a 2DOF robotic arm. The planar robotic arms have vertical movements, so the terms of earth's gravity do not appear in the dynamic equations of the system, and hence, (8.1) and (8.2) boil down to

$$M_m(q_m)\ddot{q}_m + C_m(q_m, \dot{q}_m) = F + \tau_m \qquad (8.18)$$
$$M_s(q_s)\ddot{q}_s + C_s(q_s, \dot{q}_s) = F - \tau_s \quad . \qquad (8.19)$$

For one arm with two links, we can write

$$\tau_{ij} \pm F_{kj} = \left(\frac{1}{3}m_1 l_1^2 + m_2 l_1^2 + \frac{1}{3}m_2 l_2^2 + m_2 l_1 l_2 \cos(q_{i2})\right)\ddot{\theta}_1^2$$
$$+ \left(\frac{1}{3}m_2 l_2^2 + \frac{1}{2}m_2 l_1 l_2 \cos(q_{i2})\right)\ddot{q}_{i2}^2$$
$$- \left(m_2 l_1 l_2 \sin(q_{i2})\right)\dot{q}_{i1}\dot{q}_{i2}$$

$$- \left(\frac{1}{2} m_2 l_1 l_2 sin(q_{i2}) \right) \dot{q}_{i2}^2 \tag{8.20}$$

where $i = m, s$ for master and slave, $k = e, h$ stands for the environmental and human forces that the negative sign is for F_e and positive sign for operator force Fh, and finally, $j = 1, 2$ denotes the link number. The environment task is modeled as [7]

$$F_{hj} = \begin{cases} 0 & if \ q_{sj} < d_e \\ -B_{ej} \dot{q}_{sj} - K_{ej}(q_{sj} - d_e) & if \ q_{sj} \geq d_e \end{cases} \tag{8.21}$$

where d_e is the distance angle to the obstacle, B_{ej} and K_{ej} are the damping coefficient and the stiffness coefficient of the environment, respectively, which are considered for hard motion. Furthermore, the human force F_h is considered as follows:

$$F_{hj} = K_j(q_{dj} - q_{mj}) \tag{8.22}$$

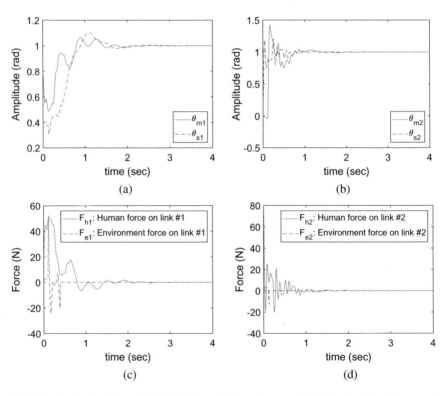

Fig. 8.3 a Position tracking of the first link in contact motion case. **b** Position tracking of the second link in contact motion case. **c** Human and environment forces in contact motion case. **d** Human and environment forces in contact motion case

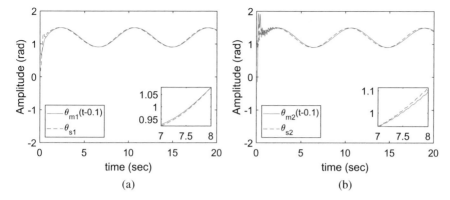

Fig. 8.4 a The synchronization performance under sinusoidal reference signal for the first joint. **b** The synchronization performance under sinusoidal reference signal for the second joint

where K_j is the stiffness coefficient and q_{dj} is the final angle to be followed by the master and slave robots, which is taken to be as equal to $q_{dj} = 1 + 0.3sin(\frac{12\pi t}{50})$. In simulations, the system parameters are chosen as follows: $m_1 = 2\,\text{kg}, m_2 = 1\,\text{kg}, l_1 = 0.4\,\text{m}, l_2 = 0.3\,\text{m}, K_1 = 100\,\text{N/m}, K_2 = 100\,\text{N/m}, B_{e1} = 10\,\text{Ns/m}, B_{e2} = 0.3\,\text{Ns/m}, K_{e1} = 30\,\text{N/m}, K_{e2} = 5\,\text{N/m}, d_e = 0.5\,\text{m}$, and the time delay set to 0.1 s. It is considered that the arm has the some contact with some obstacles in the environment.

In this case, the environment task exerts a force on the slave robot. The results are demonstrated in Fig. 8.3a–d. Figure 8.3a and b depicts the tracking performance and synchronization between the master and the slave systems with a step reference signal. Moreover, as can be seen from Fig. 8.3d, in the case of contact motion, the human force converges to the environment force and hence, complete transparency between the master and the slave systems occurs. The synchronization performances under the sinusoidal reference signal for both joints are illustrated in Fig. 8.4a and b.

8.1.4 Discussions

The controller, which is developed in this section, benefits from a stable sliding-mode adaptive fuzzy controller for bilateral teleoperation of two cooperating flexible robotic manipulators. The transparency objective is defined to include the position synchronization between the master and the slave systems. Since an imperfect communication line transmits signals between the master and the slave systems, the time delay is assumed to exist in the position transmitted from the master to the slave system. An adaptive fuzzy controller, based on sliding-mode theory, is proposed to obtain the transparency between the two systems. The proposed control scheme benefits from Mamdani-type fuzzy systems to approximate the nonlinear dynamics of the systems. A rigorous stability analysis is given for the system using an appropriate Lyapunov function. The adaptation laws for the fuzzy systems and the control signal are derived from this Lyapunov function. The required constraints and the required bounds for the selection of the parameters are obtained. The proposed control approach is simulated on flexible joint master and slave robotic arms. The simulation results show that the bilateral transparency between the master and the slave systems is achieved and the synchronization error is very low.

References

1. Van Cuong, P., Nan, W.Y.: Neural Comput. Appl. 1–12 (2015)
2. Yoo, S.G., Chong, K.T.: Neural Comput. Appl. **25**(6), 1249 (2014)
3. Ahn, H.S.: In: 2010 International Conference on Control Automation and Systems (ICCAS), pp. 1362–1365, IEEE (2010)
4. Eusebi, A., Melchiorri, C.: IEEE Trans. Robot. Autom. **14**(4), 635 (1998)
5. Nasirian, A., Khanesar, M.A.: In: 2016 International Conference on Automatic Control and Dynamic Optimization Techniques (ICACDOT), pp. 7–12, IEEE (2016)
6. Park, J.H., Cho, H.C.: In: 1999 IEEE/ASME International Conference on Advanced Intelligent Mechatronics, 1999. Proceedings, pp. 311–316, IEEE (1999)
7. Kim, B.Y., Ahn, H.S.: In: 2009 IEEE International Symposium on Computational Intelligence in Robotics and Automation (CIRA), pp. 388–393, IEEE (2009)

Chapter 9
Intelligent Optimization of Sliding-Mode Fuzzy Logic Controllers

Optimization is the selection process of the best elements with respect to some criterion from a feasible set of variables. There may be single or multiple objectives to be considered during optimization. The optimization process generally involves the minimization of a cost or maximization of a profit. Every engineering design problem ends up with the selection of values for its parameters which may include trials and errors. However, an automatic way of choosing design parameters optimally is to define a cost function and encode the decision variables such that a computer algorithm can automatically find the optimum solution to the problem.

As mentioned in previous chapters, sliding-mode control demonstrates its appropriate responses in the presence of uncertainties and complexity of nonlinear systems. In addition to fuzzy logic systems (FLSs), a number of soft computing approaches, such as heuristic optimization algorithms, may be used to handle complicated systems. One of the most common and popular methods for optimization is genetic algorithms (GAs), which works based on Darwin's theory of evolution. Darwin's theory is based on natural selection, which is adapted to machine language. Swarm intelligence, which is based on stochastic and unpredictable movement of birds, fish, and other species, is another major field of optimization which can be used for optimization purposes. Similar to any other engineering applications, sliding-mode controller design includes the selection of some parameters which need to be selected. Thus, the optimal design of sliding-mode controllers using intelligent optimization algorithms may be more interesting—when compared to the trial and error method—as it results in an automatic procedure to improve the performance of a closed-loop control system.

In [1], Chine cheng et al. used GA to increase the speed of reaching sliding mode. However, increasing the speed of reaching sliding mode increases undesirable high frequency oscillations. In order to decrease high- frequency chattering, an index term based on oscillations in the control signal is added as a cost function to be optimized.

© Springer Nature Switzerland AG 2021
M. Ahmadieh Khanesar et al., *Sliding-Mode Fuzzy Controllers*, Studies in Systems,
Decision and Control 357, https://doi.org/10.1007/978-3-030-69182-0_9

$$g_1(HT) = \left[\frac{1}{1 + \frac{HT}{\delta_1}} \right]^2$$

$$g_2(CH) = \left[\frac{1}{1 + \frac{CH}{\delta_2}} \right]^2 \tag{9.1}$$

where CH represents high-frequency chattering in the control signal and HT is the reaching time to the sliding surface. It is desired that both cost functions are minimized simultaneously to obtain a small reaching time while avoiding oscillations. In order to optimize these two objectives simultaneously, they are converted to a single-objective function by multiplying them together. However, the optimization of such a single-objective cost function may not result in the optimization of each of the cost functions separately.

Sliding-mode controllers have been used for the navigation and propulsion system of Cybership II [2]. Cybership II is a scale model of an oil platform supply ship, which is available at Marine Cybernetics Lab in the Norwegian University of Science and Technology. In the proposed structure, GA is used to optimize the parameters of the controller. The cost function is composed of tracking errors and control efforts. It has been shown that the proposed controller is capable of performing the control task with high performance under high levels of disturbances in the system.

The evolutionary algorithm has been used to find the best sliding surface to minimize the rise time and the settling time of the system [19]. The weighted sum of these two objective functions is considered as a mean to convert the multi-objective optimization algorithm to a single-objective function.

In most existing optimizations algorithms applied to tuning sliding-mode controllers, multiple objective functions are converted into a single-objective function. As an objective function, other than the integral of a squared error, some other performance indexes may be taken into the account. Examples of such performance indexes are reaching time, chattering, control effort, and the response of a system to disturbances. Since these cost functions may conflict with the cost function associated with performance, simultaneous optimization of them is truly a multi-objective optimization problem. In this chapter, in addition to single-objective optimization algorithms, multi-objective optimization algorithms are also considered.

9.1   Single-Objective Optimization Algorithms

It is a single-objective optimization problem when the objective function to be optimized is one dimensional or if there exist multiple objective functions, which are combined into a single-objective function. If the objective function is described in an explicit differentiable form, different computational methods can be used to calculate the optimum value of the objective function. Various nonlinear programming methods can be used to optimize the problems when the cost function and/or its constraints are nonlinear. Gradient descent, Newton's optimization method, Gauss–Newton algorithm, and Levenberg–Marquard are among the most commonly used nonlinear optimization methods. Most of these optimization algorithms require an explicit cost function of the variables to be known and fulfill certain requirements such as being continuous, being sufficiently smooth and being differentiable. There exist some problems that cannot be solved using mathematical optimization problems. If the objective function is not explicitly defined, most of the classical optimization algorithms fail. As an example of such problems, consider the parameter estimation of a nonlinear identifier from a finite set of sampled data. In this case, there exists no accurate relationship between identifier parameters and the sum of squared error which is to be optimized. The optimal parameter tuning of a controller is another example of optimization problems in which there exists no accurate mathematical relationship between the parameters and the cost function. In these cases, a simulation must be carried out for some time to calculate the cost function values associated with a set of parameter values suggested by the optimizer. On the other hand, there might exist a mathematical relationship between the decision parameters and the objective function but it is hard to find the optimal solutions because of the high dimension or the high number of local optimums.

When classical optimization algorithms fail to find the solution to an optimization problem, intelligent optimization algorithms may be preferred. Intelligent optimization algorithms can mainly be put into three main categories as follows

1. Evolutionary optimization algorithms,
2. Swarm optimization algorithms, and
3. Physics-inspired optimization algorithms..

9.1.1 Evolutionary Single-Objective Algorithms

9.1.1.1 Genetic Algorithm-Based Optimization Methods

Genetic algorithms (GAs) are based on natural selection and were initially introduced by Fraser [3, 4]. GAs generally result in more diverse solutions than most other intelligent optimization methods that rely on bio-inspired operators such as mutation, crossover, and selection. To use the principals of natural selection to find the optimum of a function, each optimum solution is considered to be a chromosome with its number of genes being the same as the dimension of the solution space. The values considered for each gene can be either binary value or real value, each of which has its own genetic operators. Since the parameters of a sliding-mode controller to be optimized in this chapter are real, real-valued GA is considered. Each individual may be selected as a parent of the offspring and/or survive to the next generation based on its fitness value. The crossover operator is utilized to combine two or more parents to produce an offspring. The mutation operator can be used to generate an individual from a single parent using some random operators.

Selection: There exist various kinds of selection operators in the literature. The two frequently used selection algorithms are tournament selection and roulette wheel. In the selection of the parents using a roulette wheel, the probability of selection of an individual is proportional to its fitness value. Hence, this sort of selection acts as a roulette wheel whose portions are not equal. Tournament selection is another type of selection, in which the first two or more chromosomes are selected randomly and then based on their fitness values, one of them is selected.

Crossover operator: There exist various crossover operators for real-valued GA, which can be applied on two or more parents chosen using the selection operator to produce offsprings. The heuristic crossover operator, which was developed by Wright, is as follows [5]:

$$Offspring_{ij}(t) = U(0, 1)\big(Parent_{2j}(t) - Parent_{1j}(t)\big) + Parent_{2j}(t)\big) \quad (9.2)$$

where the fitness function of $Parent_2$ is better than that of $Parent_1$. This crossover operator is a convex one and the produced offspring falls between the two parents. The multi-parent version of this convex crossover operator is as follows:

$$Offspring_{ij}(t) = \sum_{i=1}^{n} a_i Parent_{ij}(t) \quad (9.3)$$

where $a_i \in [0, 1]$ and $\sum_{i=1}^{n} a_i = 1$. However, it is possible to use a non-convex crossover operator that shares the same equation for crossover as in (9.3), in which, the constraints of a_i's are different and satisfy the following equations:

$$a_i \in [-0.5, 1.5] \ and \sum_{i=1}^{n} a_i = 1. \quad (9.4)$$

The use of a non-convex crossover operator increases the exploration of the search space.

Mutation: This operator results in the better exploration of the solution space and avoids the entrapment in a local optimum. There exist different real-valued mutation operators. A typical mutation operator is as follows:

$$Offspring_{ij} = \begin{cases} Parent_{ij} + U(0, 1)\big(MaxVal_j - Parent_{ij}\big) \ if \ r = 0 \\ Parent_{ij} + U(0, 1)\big(Parent_{ij} - MinVal_j\big) \ if \ r = 1 \end{cases} \quad (9.5)$$

where r is a random digit selected as $r \in [0, 1]$, $MaxVal_j$ is the maximum value in the jth dimension, and $MinVal_j$ is the minimum value in the jth dimension. The general flowchart of GA is illustrated in Fig. 9.1.

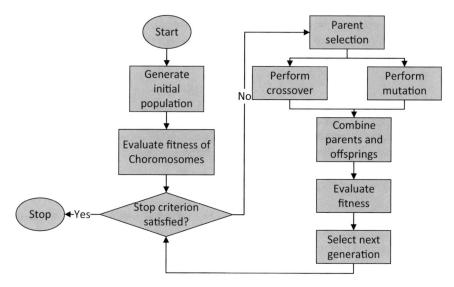

Fig. 9.1 General flowchart of GA

9.1.2 Swarm Intelligence

9.1.2.1 Particle Swarm Optimization

The PSO algorithm is a population-based search algorithm that relies on the motion behavior of birds or a school of fish. Although the motion behavior of a bird among folks seems to be completely unpredictable, it exhibits certain rules such as leader following, replacement of leader, and avoiding undesirable positions. These rules make it possible for the swarm to fly synchronously and efficiently to explore new positions. Sudden changes in flight direction due to an invader or a new target is also observed in the behavior of birds, which make their motion behavior a suitable method to be used as an optimization algorithm.

The dimension of position associated with particles within PSO is equal to the dimension of the solution space. The positions of the particles changes on the basis of a velocity vector, which is influenced by self-experiences and social experiences. The social experience may be considered to be the general experience of the whole swarm or its close neighbors. An inertia term is considered to maintain the last motion direction.

In order to use PSO for the optimization of a cost function, it is required to encode the solution space to the position of particles. Unlike GA which is basically based on binary values, PSO is mainly designed for real-valued optimization problems. However, this algorithm is later extended to binary-valued optimization problems as well.

A d-dimensional solution is modeled as the position of the swarm represented by x_i, in which, i counts the number of particle within the swarm. The position vector is updated using a velocity vector v_i as follows:

$$x_i(t+1) = x_i(t) + v_i(t+1) \qquad (9.6)$$

where $x_i(t)$ belongs to the interval $[x_{min}, x_{max}]$. The velocity vector v_i is responsible for exchanging information between particles within the swarm. There exist three different terms in the update equation of a particle as follows:

$$v_i(t+1) = wv_i(t) + r_1c_1(x_{Pbest,i} - x_i(t)) + r_2c_2(x_{sbest} - x_i(t)) \qquad (9.7)$$

where x_{Pbest} is the recorded best position observed by a particle, x_{sbest} represents the social experience of particles observed by other members of the swarm and wv_i is the inertia term that maintains the last movement direction.

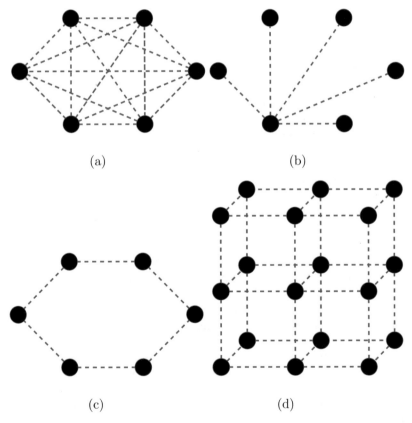

Fig. 9.2 Various neighborhood structures: **a** star neighborhood; **b** ring structure; **c** wheel structure; and **d** Von Neumann

The x_{sbest}, which records the social experience of the swarm, may be updated based on the best position observed by all members of the swarm or neighboring particles.

It is a global PSO algorithm, if the size of the neighborhood is considered to be exactly the same as the size of the swarm. In this case, the exchange of information is done within the whole swarm. Hence, all particles in a swarm move toward the best experience found by one individual. This may cause fast convergence of individuals to the best position found. Too fast convergence is not desired as the best experience found may be far from the global optimum of the function, especially in the first few iterations. To avoid fast convergence, it is possible to exchange information within a neighborhood.

There exist different neighborhood structures, examples of which are star neighborhood, ring structure, wheel structure, and Von Neumann. Figure 9.2 illustrates these neighborhood policies, in which each particle can only transmit data with the particle connected by a dashed line. In a star-connected PSO, all particles become aware of the best solution found as it is observed by any particle. However, in other neighborhood strategies, the best solution is propagated slowly giving rise to more exploration in the search space.

9.2 Multi-objective Optimization

In single-objective optimization problems, the ultimate goal is to find the best—or the most optimal—solution, which is called the global optima. However, in some optimization problems, there is not just one objective function but several objective functions need to be optimized. It frequently happens that the objective functions are incompatible with each other; thus there are some trade-offs between them. These kinds of problems with two or more objective functions are called *multi-objective optimization problems*. To solve multi-objective optimization problems, single-objective optimization algorithms must be modified considerably. In the case of multi-objective optimization problems, new definitions such as dominance, Pareto front, and new solution strategies such as repository may be used.

The term "optimality" alters when interacting with multi-objective optimization problems. Metaheuristics are more popular and successful strategies for most of the optimization problems. Among different metaheuristic approaches that have been emerged over the last two decades, evolutionary algorithms provide a couple of advantages to solve optimization problems. These advantages include achieving robust performance, enabling global search, requiring no derivative evaluation of objective functions and the constraints, and having little or no information about the problem. Moreover, due to the population-based nature of evolutionary algorithms (EAs), the utilization of them in both single- and multi-objective problems is an appropriate choice.

Let X be an n-dimensional feasible search space, $f_i(x)$, $i = 1, \ldots, M$ introducing an M objective function over decision vector x and $g_i(x) \leq 0$, $i = 1, \ldots, K$ be K

inequality constraints. Multi-objective optimization problems can be described as
specifying a vector $x = (x_1, x_2, \ldots, x_m) \in X$ as a solution while satisfying the
imposed constraints. The objective function in vector form is as follows:

$$F(x) = (f_1(x), \ f_2(x), \ \ldots, \ f_M(x))^T. \tag{9.8}$$

In the context of multi-objective optimization, the Pareto optimality concept is
introduced. A decision vector is determined in the Pareto optimal solution or non-
dominated solution if no solution dominates it. For a minimization problem, the
solution x_A dominates solution x_B if

$$f_1(x_A) \le f_1(x_B), \ \ldots, \ f_M(x_A) \le f_M(x_B). \tag{9.9}$$

The set of all the Pareto optimal solutions construct the Pareto optimal set. Therefore, the concept of quality of solutions is more sophisticated and may contain more
solutions in multi-objective optimization problems than its single-objective function
counterpart.

9.2.1 Elitist Non-dominated Sorting Genetic Algorithm (NSGA-II)

In the last two decades, various multi-objective evolutionary algorithms (MOEAs)
have been studied to deal with MOPs in the literature. As mentioned earlier, the main
motivation is their potential to find a set of Pareto optimal solutions in one run. One
of the first and most effective MOEAs is non-dominated sorting genetic algorithm
(NSGA) proposed by Srinivas and Deb [6]. Despite its benefits, the main criticisms
are as follows:

1. High computational complexity of non-dominated sorting: NSGA has the computational complexity of $O(MN^3)$ (where M is the number of objectives and N is
   the population size), which is extremely high when considering this algorithm for
   large population sizes. The non-dominated sorting procedure in each generation
   is the main cause of the high complexity of this algorithm.
2. Lack of elitism: It is clear that elitism can boost the convergence speed of the GA.
   It also helps to avoid the lack of good solutions after they have been found.
3. Need for specifying the sharing parameter σ_{share}: In order to obtain a high diversity
   in the population and subsequently, generate a variety of high-quality solutions,
   NSGA applied the sharing parameter σ_{share}. However, the specification of the
   sharing parameter is one of the bottlenecks of the algorithm.

Elitist non-dominated sorting genetic algorithm-II (NSGA-II), an improved version
of NSGA, is proposed by Deb et al. [7]. In order to alleviate all the difficulties of the
NSGA, the NSGA-II has applied the following mechanisms:

1. *Non-dominated sorting:* Non-dominated sorting is a mechanism for assigning a value to each solution based on the front it belongs to. To identify the solutions that are not dominated by other solutions, they need to be compared against all other solutions in the population. These solutions compose the first non dominated front. When this mechanism is continued, all solutions in the first non-dominated front are found. At the next stage, the solutions within the next non-dominated front will be discovered and the solutions of the first front are discounted temporarily. The above mechanism is repeated until the following fronts are found.

2. *Density Estimation:* Along with convergence to the Pareto optimal set, it is also desired that an MOEA maintains a good spread of solutions in the obtained set, thereby requiring a measurement for diversity. The average distance of the two solutions on either side of a specific solution along each of the objectives is the crowding distance criteria. This quantity is considered as an estimation of the density of solutions surrounding a particular solution in the population. These computations can easily be implemented by sorting the population with respect to each objective function value in ascending order. After that, the solution given by the boundary of the sorted set for each objective function is given an infinite distance value. The reminder solutions' distance values are calculated with respect to their two adjacent solutions. The procedure of diversity measurements does not have any user-defined parameters.

3. *Crowded-Comparison Operator:* With the two above-mentioned mechanisms, each solution of the population has two attributes: non-domination ranking and crowding distance. With the aim of approximating the true Pareto optimal front, the crowded-comparison operator selects the non-dominated solutions that are not in a congested region of the Pareto front.

The pseudocode for NSGA-II is as in Table 9.1.

Table 9.1 Pseudo code for NSGA-II

Initialize N number of population with d genes for each of them
For iteration = 1 to Maximum number of generations
Generate offspring
Select parents
Perform crossover operation on selected parents
Perform mutation operation
Merge parents with offspring to generate a new population size equal to 2N
For each individual= 1 to 2N
Calculate the Pareto front number based on non-dominated sorting
Select the next generation using the rank of each chromosome
For chromosomes in the last Pareto front based on crowding distance
End
End

9.3   Strength Pareto Evolutionary Algorithm (SPEA2)

Strength Pareto Evolutionary Algorithm 2 (SPEA2), an improved version of SPEA [8], was introduced by Zitzler and Thiele [9]. SPEA has its deficiencies and weaknesses. The list below describes these aspects:

1. *Fitness assignment:* The procedure of fitness assignment to each solution in SPEA involves only the archive of non-dominated solutions. Subsequently, the selection ability of the SPEA algorithm decreases as the case when members of the population dominating each other are not taken into account.
2. *Density estimation:* This technique is only done on the archive but not on the population. When solutions do not have any dominance relation with each other, there are no established criteria to determine better solutions in the population.
3. *Archive truncation:* SPEA uses a clustering technique to remove extra non-dominated solutions of the excessive archive. However, these center-based techniques may not be appropriate by losing boundary solutions and does not result in extending the Pareto optimal front.

SPEA2 has been designed to address the issues above. For this purpose, SPEA2 applies three well-defined strategies, including an improved fitness assignment, a nearest neighbor density estimation technique, and an enhanced archive truncation method. The improved fitness assignment provides density information but is not useful in the case when solutions do not dominate each other so that additional density information is required. The second strategy uses an adaptation of the kth nearest neighbor method to find the density of each solution. Finally, the third strategy determines which solution should be put out of the archive by developing two substrategies: (1) the number of elements in the archive is kept constant throughout generation and (2) the truncation method does not guarantee the removal of the boundary solutions and hence, leads to well-spread solutions.

The pseudocode describing the overall SPEA2 algorithm is presented in Table 9.2.

9.3.1   Multi-objective Particle Swarm Optimization (MOPSO)

Particle swarm optimization is a heuristic optimization method motivated by the observing behavior of birds within a flock. The intrinsic simplicity and high speed of convergence indicated when applied to single-objective optimizations made it well suited to be extended for multi-objective optimization, which is called multi-objective particle swarm optimization. In a MOPSO algorithm, three important ingredients are developed to incorporate the multi-objective approach into PSO:

1. Altering the updating rule of PSO in order to keep a balance between convergence and diversity.

Table 9.2 Pseudocode of the overall SPEA2 algorithm

1. Initialize a population P_t, t is time step, and create the initial external archive, P_0' of size N';
2. Calculate fitness values for individuals in P_t and P_t' and a strength value $S(i)$ is assigned to individual i as shown in (9.10):

$$S(i) = |\{j \,|\, j \epsilon P_t + P_t' \wedge i \succ j\}| \tag{9.10}$$

(2-1) A raw fitness is calculated for both the archive and population set as follows:

$$R(i) = \sum_{j \epsilon P_t + P_t', j > i} S(j) \tag{9.11}$$

(2-2) Use density estimation technique (inverse distance of the kth nearest neighbour), k is the square root of population size plus archive size:

$$D(i) = \frac{1}{\sigma_i^k + 2} \tag{9.12}$$

(2-3) Add the density metric to the raw fitness:

$$F(i) = R(i) + D(i) \tag{9.13}$$

3. Copy all non-dominated individuals in P_t and P_t' to P_{t+1}'
   (3-1) If $|P_{t+1}'| > N'$, Employ truncation procedure that iteratively removes individuals from P_{t+1}' until $|P_{t+1}'| = N'$ according to (9.14) and (9.15):

$$i \leq_d j \ for \ all \ j \in P_{t+1} \tag{9.14}$$

$$i \leq_d j :\Leftrightarrow \forall 0 < k < |P_{t+1}'| : \sigma_i^k = \sigma_j^k \ \vee \tag{9.15}$$
$$\exists 0 < k < |P_{t+1}'| : [(\forall 0 < l < k : \sigma_i^l = \sigma_j^l) \wedge \sigma_i^k < \sigma^{k_j}]$$

   (3-1) else Copy the best $N' - |P_{t+1}'|$ dominated individuals in the previous archive and population to the new archive
4. Apply mating selection by binary tournament selection with replacement on P_{t+1}'.
5. Stop if the termination is satisfied and Return non-dominated individuals in P_{t+1}' . Otherwise, turn back to Step 2.

---

2. Applying a contraction and expansion strategy to adopt the size of objective space to reach the true optimal Pareto front.
3. Developing diversity by positioning new particles in the sparse area and eliminating particles in populated regions.

Table 9.3 General pseudocode for MOPSO

1. Initialize swarm population and velocity.
2. Fitness evaluation and Pareto dominance for ranking particles.
3. Store personal best in the memory.
4. Store non-dominated solutions in external archive.
5. Particles select global leaders from external archive.
6. Update the position and velocity of particles by the following equation.
7. Fitness evaluation and Pareto dominance for ranking particles.
8. Apply Mutation.
9. Update Personal Best.
10. Maintain external archive.
11. Go to STEP 5 if stopping criteria is not satisfied.
12. Report solutions from external archive.

As opposed to PSO where the global best particle is determined once for each generation, MOPSO uses a set of different leaders—called the external archive—in which every particle can select a random element of it. At the end of the optimization procedure, the members of the external archive are reported as a representative set of Pareto optimal solutions. Most common MOPSO uses a mutation operator that encourages not only diversity in the objective space but also a full search of the decision space. The updating procedure of particles' personal best accomplishes according to the dominance relationship between the new particle and its personal best. Furthermore, in contrast to most MOEAs, MOPSO applies the region-based selection strategy rather than an individual one. It is to be noted that MOPSO has a low computation complexity and converges reasonably fast. A general pseudocode for MOPSO is presented in Table 9.3.

9.4 Multi-objective Optimization of Sliding-Mode Fuzzy Logic Controllers

One of the challenges met by sliding-mode fuzzy logic controllers is that it is difficult to tune its parameters optimally. The parameters, which may need to be determined during the design procedure, can be categorized as follows:

1. Coefficients of the sliding surface;
2. Parameters of the Signum function, which can be its gain, and possibly other parameters that may be used to smoothen the Signum function;
3. Parameters of the identifiers used to identify the unknown dynamics of the system.

Although an appropriate Lyapunov function can be used to prove the system's stability, the choice of the aforementioned parameters may be a difficult task and may

Fig. 9.3 Structure of the modified controller

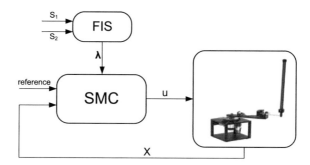

require several trials and errors. There exist no analytical optimal solutions to select parameters of a sliding-mode controller. Hence, intelligent optimization algorithms may be used to tune the parameters of the system. Other than performance indexes related to the energy of error and its time derivative both of which are frequently used in optimization, an index indicating the level of chattering may also be used. While the former is used frequently, the latter cost function solely belongs to sliding-mode controllers. Considering the fact that high-performance control of this system in the presence of unmodeled dynamics may need high-frequency switching, the cost functions based on performance indexes and chattering are the two conflicting cost functions, the optimization problem is a truly multi-objective optimization problems.

According to [10], there are several obstacles affecting the widespread applications of SMC such as chattering, matched and unmatched uncertainties, and unmodeled dynamics. Hence, the integration of SMC, soft computing and other technologies made the implementation of controllers much easier. In this chapter, an adaptive sliding-mode fuzzy controller is designed for the rotary inverted pendulum, in which, the FLS is used to estimate the equivalent control and the switching control. Moreover, control parameter selection and model parameter estimation in the design of SMC have a big impact on chattering in the control signal and can influence robustness. Hence, the motivation is to use MOEAs to tune these parameters based on desired objectives. The structure of the proposed controller is as shown in Fig. 9.3. Various existing MOEAs are utilized to tune the parameters of FLS. The existing adaptation improves the performance of the system and reduces the chattering index.

9.4.1 Sliding-Mode Fuzzy Controller for Rotary Inverted Pendulum

The rotary inverted pendulum is a widely investigated nonlinear system due to its static instability. A version of this system is used as a simplified model for the control of rider-motorcycle systems in circular motion on paths of different radii [11]. The state-space representation of the rotary inverted pendulum is as follows:

$$\dot{x}_1 = x_2 \tag{9.16}$$
$$\dot{x}_2 = F_1(X) + G_1(X)u$$
$$\dot{x}_3 = x_4$$
$$\dot{x}_4 = F_2(X) + G_2(X)u$$

where

$$F_1(X) = \frac{-5.472 \times 10^{-2} \sin(2x_1) x_2^2}{0.31425 - 0.10944 \cos^2(x_1)} - \tag{9.17}$$
$$\frac{2.428x_4 \cos(x_1) - 14\sin(x_1)}{0.31425 - 0.10944 \cos^2(x_1)}$$

$$G_1(X) = \frac{0.0911}{3.1425 \times 10^{-5} - 1.0944 \times 10^{-5} \cos^2(x_1)} \cos(x_1)$$

$$F_2(X) = \frac{0.0047 \, F_1(X) - 0.2054\sin(x_1)}{0.0033\cos(x_1)}$$

$$G_2(X) = \frac{0.1289}{0.0067 - 0.0023 \cos^2(x_1)}.$$

As the nonlinear zero dynamics of this system are unstable, this system is non-minimum phased [12], which makes the design process of the SMC more complicated. Consider a system in strict feedback form as follows:

$$x^{(n)} = F(x) + G(x)u$$
$$y = x. \tag{9.18}$$

In order to control this non-minimum phase system, the mathematical model of the system can be seen as a combination of two systems in a canonical form. Consider desired dynamics for these two canonical forms:

$$s_1 = \dot{x}_1 + \lambda_1 x_1 \tag{9.19}$$
$$s_2 = \dot{x}_3 + \lambda_3 x_3.$$

These two equations define the dynamic behavior of the system on the sliding surface. Considering these sliding surfaces, a candidate Lyapunov function can be proposed as follows:

$$V = |s_1| + \lambda |s_2|. \tag{9.20}$$

Calculate the derivative of V along the trajectories of the system:

$$\dot{V} = -\alpha \, sat(V) \tag{9.21}$$

The following control signal is obtained:

$$u = \{-\alpha sat\,(V) - (\lambda_1 x_2 + F_1\,(X))\,sgn\,(s_1)$$
$$-\lambda\,(\lambda_3 x_4 F_2\,(X))\,sgn(s_2)\} \times$$
$$\{G_1\,(X)\,sgn\,(s_1) + \lambda G_2\,(X)\,sgn(s_2)\}^{-1}. \tag{9.22}$$

The control signal of (9.22) stabilizes the system but chattering in the control signal may occur.

The parameter λ is a very important parameter, which has a significant influence on the performance of the controller. In order to increase the degrees of freedom of the controller, this parameter is considered to be the output of an FLS. The basic idea of this modification is that when the pendulum is in its upright position, it is suggested that the control algorithm takes more care of the position of the motor. However, when the pendulum is far from its origin, it is more recommended that λ takes a smaller value, so that the control algorithm takes more care of the position of the controller.

The inputs of the FLS are s_1 and s_2. The multi-objective optimization algorithm optimizes the parameters of the FLS including the centers of membership functions and their sigma values as well as consequent part parameters. The number of parameters of the FLSs that are to be optimized in this technique is equal to 25.

9.4.2 Multi-objective Tuning of the Parameters of Sliding-Mode Fuzzy Logic Controller

Rather than tuning the parameters of SMC using the trial-and-error method, evolutionary algorithms are utilized to tune these parameters based on desired objectives. In [13], Chin Chang et al. used GA to optimally tune the parameters of the sliding-mode controller. The objective function considered in their research was the minimization of reaching time of the sliding-mode controller. They suggested that the minimization of the reaching time increases the chattering in the control signal; consequently, chattering is used as the second objectives. By using multiplication, these two objective functions are converted to one objective function. They used GA to optimize the parameters of the sliding-mode controller. In [14], the authors used EA to optimize two objectives of the rise time and the settling time of the system simultaneously. In many controller design problems, these two objective functions are conflicting with each other. The optimization of settling time makes the system faster, which, in turn, may destroy the rise time of the system. In addition, an FLS is used to convert the multi-objective optimization problem to a single-objective one, which is optimized using GA. Similarly, in [14] the multi-objective design process of the SMC is treated as a single-objective problem. The optimal values are obtained using EA.

In this chapter, the optimization is done based on two objectives. The first objective relates to the performance of the system and the second one is an index associated with chattering. Based on the technical specification of the system and the Lyapunov function, some constraints are introduced. Different MOEAs are iterated for a sufficient number of epochs and some comparisons are made using these MOEAs.

The parameters to be optimized in the proposed fuzzy controller structure are as follows:

1. Parameters of the sliding surface λ_1, λ_2, λ_3 and the parameter of the Lyapunov function α;
2. Number of parameters for the premise parts of the rules (center and sigma values of the Gaussian MFs);
3. Number of parameters for consequent parts of the rules.

The objective functions considered in this study for MOPs are the measurement of the system performance and the chattering index.

There are some well-known performance indexes based on the control error. Since the rotary inverted pendulum output is the angle of the pendulum (x_1),

$$e = x_{1d} - x_1 = -x_1 \tag{9.23}$$

in which x_{1d} is the reference signal, which is considered as being equal to *zero* in this chapter. A performance index for the system may be introduced as follows:

$$\int_0^\infty e^2(t)dt \tag{9.24}$$

which is called the integral of squared error. When an error is near *zero*, its square is smaller than the real value of the system and when it is bigger than *one*, its square is bigger than its real value. Therefore, the controller tuned based on this performance index shows a high reaction to big errors and a weak reaction to small ones. These controllers may suffer from highly oscillating responses. In addition, the minimization of this performance index results in the minimization of power of the error [15]. Since it is desired to optimize the response of the controller considering the integral of squared error and its time derivative, a performance index is considered as follows:

$$\int_0^\infty \left[e^2(t) + \dot{e}^2(t) \right] dt. \tag{9.25}$$

In order to optimize the second objective function, which is chattering, a quantitative formula is required. There exist various chattering indexes in the literature; the following chattering index can be mentioned as the first example [16]:

$$C = \frac{\int_T \left| \frac{d^2 u}{dt^2} \right| dt}{u_{max}}. \tag{9.26}$$

Equation (9.26) uses the average absolute second-order derivative value of the actuator output during time interval T divided by the maximum actuator output as a chattering index. This definition reflects the idea that the degree of chattering can approximately be estimated as the average change rate of the slope of the actuator output. In [17], Erbatur and Kawamura introduced a chattering index as the absolute derivative of the error and used this performance index to automatically tune the boundary value of the sliding-mode controller. The chattering index they introduced is as follows:

$$C = |\dot{u}| . \tag{9.27}$$

Using a low-pass filter, the oscillation of the control signal can be eliminated; consequently, the difference between the filtered signal and the main signal can be accepted as a measurement for chattering in the control signal as follows:

$$C = |u - u_f| , \tag{9.28}$$

where u_f is the filtered control signal, which is defined as follows:

$$\tau \dot{u}_f = u - u_f , \tag{9.29}$$

where τ is a user-defined parameter. Among the introduced methods for chattering, the second one is the simplest one. In this chapter, (9.29) is used as a measurement of chattering. Since the accumulated value of such chattering measure is required to be optimized, an integral is added to the function as follows:

$$\int_{t=0}^{T} |u - u_f| dt . \tag{9.30}$$

Some constraints are considered in the proposed optimized fuzzy sliding-mode controller. The first constraint for the optimization is imposed by the fact that the Lyapunov function is a positive-definite function and the parameter α needs to be positive. In addition, parameters λ_1, λ_2, and λ_3 must be positive in order to guarantee the stability of the system. In order to avoid any damage to the motor in the system, its input voltage must be restricted to 6 V, i.e., $|u| \leq 6$. To avoid the non-stop movement of the rotor, it is required to restrict the motor angle rotation as follows:

$$0 < x_3 = \theta \leq 2\pi . \tag{9.31}$$

9.4.3 Parameters Used in MOEAs

The conceptual framework for parameter tuning of different evolutionary algorithms is presented in [18]. The population size and number of generations for different MOEAs used in this manuscript are taken to be equal to 100. Moreover, the dimension of the search space is 25. The parameters considered for multi-objective optimization algorithms are considered to be as follows:

NSGA-II: The uniform crossover rate and mutation rate are taken as 0.7 and 0.3, respectively.

MOPSO: The repository size is taken as 100. The inertial weight w is taken as 0.5 and is decreased linearly by the *damping factor* $= 0.99$. The learning factors $C_1 = 1$ and $C_2 = 0.5$. Number of grids per each dimension is 10.

SPEA2: The type of crossover is chosen to be uniform. The crossover and mutation rates are taken to be equal to 0.7 and 0.4, respectively. The archive size is fixed to 100.

9.4.4   Performance Metrics

Several metrics based on the obtained non-dominated solutions are introduced to measure the searching quality of the algorithm, which is used for comparison between different algorithms.

Coverage Metric (CM)   In [19], CM is used to compare two non-dominated solution sets E and E' obtained by two algorithms, which maps the ordered pair (E, E') to the interval [0, 1] to return the dominance relationship between solutions in the two sets as follows:

$$C\left(E, E'\right) = \left|\left\{x' \in E' | \exists : x \in E, x' \prec x\right\}\right| / \left|E'\right|. \tag{9.32}$$

If all the solutions in E' are dominated by those in E, then $C(E, E') = 1$. Conversely, if all the solutions in E are dominated by those in E', then $C(E, E') = 0$. Since there may be some solutions in E and E' that are not dominated with respect to each other, the sum of $C(E, E')$ and $C(E, E')$ is not always equal to 1.

Distance Metrics (D_{av} and D_{max})   In [20, 21], two distance metrics are used to measure the performance of the non-dominated solution set E relative to a reference set R of the optimal Pareto front

$$D_{av} = \sum_{x_R \in R} \min_{x \in R} d(x, x_R)/|R| \tag{9.33}$$

$$D_{max} = \max_{x_R \in R} \left\{\min_{x \in E} d(x, x_R)\right\}, \tag{9.34}$$

where $d(x, x_R) = \max_{j=1,\dots,n} \left\{\left(f_j(x) - f_j(x_R)\right)/\triangle_j\right\}$, $x \in E$, $x_R \in R$, and $\triangle_j$ is the range of f_j values among all the solutions in set E and set R. D_{av} is the average distance from a solution $x_R \in R$ to its closest solution in E, and D_{max} represents the maximum value of the minimum distance from a solution $x_R \in R$ to any solution in E. Obviously, smaller D_{av} and D_{max} values correspond to a better approximation to the optimal Pareto front. When comparing the metric values of two non-dominated solution sets E and E' if the optimal Pareto front is not known, the combination of the two sets is considered and all the non-dominated solutions to form set R are selected.

Tan's Spacing (TS)   In [22], the following spacing metric is used to measure how evenly the solutions are distributed:

$$TS = \sqrt{\frac{1}{|E|} \sum_{i=1}^{|E|} (D_i - \overline{D})^2 / \overline{D}} \qquad (9.35)$$

where $\overline{D} = \sum_{i=1}^{|E|} D_i / |E|$. $|E|$ is the number of solutions in the set E, and D is the Euclidean distance in objective space between the solution i and its nearest solution in the set E. The smaller the metric is, the more uniformly the solutions distribute.

Maximum Spread (MS)   This metric was defined in [22] to measure how well the true Pareto front is covered by the obtained non-dominated solutions in the set E through the hyberboxes formed by the extreme function values observed in the optimal Pareto front and E,

$$MS = \sqrt{\frac{1}{n} \sum_{j=1}^{n} \left(\frac{max_{i=1}^{|E|} f_j(x_i) - min_{i=1}^{|E|} f_j(x_i)}{F_j^{max} - F_j^{min}} \right)^2} \qquad (9.36)$$

where $|E|$ is the number of solutions in E, $f_j(x_i)$ is the jth objective of solution x_i, and F_j^{max} and F_j^{min} are the maximum and minimum values, respectively, of the jth objective in the optimal Pareto front. MS can imply abundant information. If the final Pareto optimal set E converges to the true Pareto optimal front well, for minimization problems, the MS value ranges from 0 to 1. The final solutions will distribute better over the Pareto front, if the MS value is closer to 1. Alternatively, if the MS value is equal to 0, it can be concluded that only one non-dominated solution is found. Moreover, if the final Pareto optimal set E is not around the true Pareto optimal front, the MS value will be greater than 1. Consequently, the MS value can specify the coverage range of the final Pareto optimal set E when it lies in the closed unit interval, and it can also imply the adverse convergence of the MOEA when it is greater than 1.

9.4.5 Simulation Results

In the following experiment, three algorithms are compared and some analysis is given. The objective functions are selected as in (9.25) and (9.29). In addition, the previously introduced constrains of Sect. 9.4.2 are used. After obtaining the Pareto optimal front, the designer can choose one of the solutions by using the higher level knowledge. Since there are just 100 solutions, the designer can focus on the less number of solutions and choose one solution from the obtained set. Three algorithms are simulated 20 times independently for two different numbers of function evaluations;

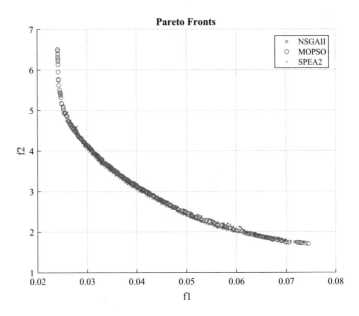

Fig. 9.4 Non-dominated front of 20 approximate Pareto fronts for different MOEAs when NFE is equal to 10000

consequently, 20 approximate Pareto fronts are obtained. Number of fitness evaluation (NFE) is a metric for comparing the complexity efficiency of each algorithm. In this chapter, the performance of algorithms are compared with two different NFE values equal to 10000 and 20000. In order to compare different algorithms intuitively, four metrics, coverage metric, distance metric, spacing metric, and maximum spread metric are used.

The non-dominated front obtained for the three optimization methods is plotted in one figure (Fig. 9.4). In Fig. 9.4, it is shown that the Pareto front MOPSO mostly dominates other Pareto fronts of the other two algorithms. Moreover, the Pareto front obtained by MOPSO is more diverse than the other two optimization algorithms.

As can be seen from Table 9.4, the mean value of MS for MOPSO when NFE is equal to 10000 is greater than that of other algorithms as it is closer to 1. The performance of MOPSO, when NFE is equal to 10000, is better than those of their counterparts considering Spacing, Maximum spread, D_{av}, and D_{max} matrices as well. For the case when NFE equals 20000 when considering the spacing metric, NSGA-II is the best one. For the other three metrics, MOPSO outperforms the other algorithms.

As can be seen from Table 9.5, the results of comparisons of the coverage metric performance of different MOEAs, in the case of NFE equal to 10000, show that MOPSO performs better than the other algorithms. However, when NFE is equal to 20000, the performance of NSGA-II is better than those of the other algorithms.

Table 9.4 The performance evaluation metrics for different MOEAs over 20 runs

NFE	Metrics	NSGA-II	MOPSO	SPEA2
10000	TS	0.2627	**0.1405**	0.1711
	MS	0.7215	**0.8636**	0.5435
	D_{av}	0.0398	**0.0077**	0.1106
	D_{max}	0.5599	**0.0604**	0.2981
20000	TS	**0.0919**	0.1499	0.2304
	MS	0.6434	**0.9329**	0.8755
	D_{av}	0.0676	**0.0073**	0.5785
	D_{max}	0.1593	**0.0685**	1.0899

Table 9.5 The results obtained for C metric for different MOEAs over 20 runs

NFE		NSGA-II	MOPSO	SPEA2
10000	NSGA-II	–	0.1886	0.2649
	MOPSO	**0.6867**	–	**0.4801**
	SPEA2	0.3884	0.2525	–
20000	NSGA-II	–	**0.3967**	**0.5802**
	MOPSO	0.3076	–	0.5264
	SPEA2	0.1314	0.1186	–

Controller design techniques involve the tuning of many free parameters that based on design criteria. Since the design of an optimal controller requires the optimization of some competitive objectives, a multi-objective algorithm is used for tuning. In this chapter, NSGA-II, MOPSO, and SPEA2 are used to obtain the parameter set for the SMC of the rotary inverted pendulum. Simulation results show the trade-off between two objective functions considered for the sliding-mode controller, which are a performance index composed of error and its derivative plus the second objective function, which is an index of chattering. Simulation results demonstrate that when NFE equals 10000, MOPSO shows better performance with respect to coverage metrics and distribution property. However, the performances of NSGA-II are better than those of its counterparts by increasing NFE to 20000.

Acknowledgements The authors would like to acknowledgement Dr. Bibi Elham Fallah Tafti for her contribution in writing the source code for the optimization made in this chapter.

References

1. Chen, C.-L., Chang, M.-H.: Optimal design of fuzzy sliding-mode control: a comparative study. Fuzzy Sets Syst. **93**(1), 37–48 (1998)

2. Alfaro-Cid, E., McGookin, E., Murray-Smith, D., Fossen, T.: Genetic algorithms optimisation of decoupled sliding mode controllers: simulated and real results. Control Eng. Pract. **13**(6), 739–748 (2005)
3. Fraser, A.S.: Simulation of genetic systems by automatic digital computers i. introduction. Aust. J. Biolog. Sci. **10**(4), 484–491 (1957)
4. Fraser, A.S.: Simulation of genetic systems by automatic digital computers ii. effects of linkage on rates of advance under selection. Aust. J. Biolog. Sci. **10**(4), 492–500 (1957)
5. Wright, A.H., et al.: Genetic algorithms for real parameter optimization. Found. Gen. Algorithms **1**, 205–218 (1991)
6. Srinivas, N., Deb, K.: Muiltiobjective optimization using nondominated sorting in genetic algorithms. Evol. Comput. **2**(3), 221–248 (1994)
7. Deb, K., Agrawal, S., Pratap, A., Meyarivan, T.: A fast elitist non-dominated sorting genetic algorithm for multi-objective optimization: Nsga-ii. In: International Conference on Parallel Problem Solving from Nature, pp. 849–858. Springer (2000)
8. Zitzler, E., Thiele, L.: Multiobjective evolutionary algorithms: a comparative case study and the strength pareto approach. IEEE Trans. Evol. Comput. **3**(4), 257–271 (1999)
9. Zitzler, E., Laumanns, M., Thiele, L.: Spea2: Improving the strength pareto evolutionary algorithm. TIK-report, vol. 103 (2001)
10. Yu, X., Kaynak, O.: Sliding-mode control with soft computing: a survey. IEEE Trans. Ind. Electr. **56**(9), 3275–3285 (2009)
11. Chen, C.-K., Lin, C.-J., Yao, L.-C.: Input-state linearization of a rotary inverted pendulum. Asian J. Control **6**(1), 130–135 (2004)
12. Khanesar, M.A., Teshnehlab, M., Shoorehdeli, M.A.: Sliding mode control of rotary inverted pendulum. In: Mediterranean Conference on Control & Automation, 2007, MED'07, pp. 1–6, IEEE (2007)
13. Wong, C.-C., Chang, S.-Y.: Parameter selection in the sliding mode control design using genetic algorithms. Tamkang J. Sci. Eng. **1**, 115–122 (1998)
14. Trebi-Ollennu, A., White, B.: Multiobjective fuzzy genetic algorithm optimisation approach to nonlinear control system design. IEE Proceed.-Control Theory Appl. **144**(2), 137–142 (1997)
15. Paraskevopoulos, P.: Modern Control Engineering. CRC Press, New York (2001)
16. Ryu, S.-H., Park, J.-H.: Auto-tuning of sliding mode control parameters using fuzzy logic. In: American Control Conference, 2001. Proceedings of the 2001, vol. 1, pp. 618–623. IEEE (2001)
17. Erbatur, K., Kawamura, A.: Chattering elimination via fuzzy boundary layer tuning. In: IECON 02 [Industrial Electronics Society, IEEE: 28th Annual Conference of the], vol. 3, pp. 2131–2136. IEEE (2002)
18. Eiben, A.E., Smit, S.K.: Parameter tuning for configuring and analyzing evolutionary algorithms. Swarm Evol. Comput. **1**(1), 19–31 (2011)
19. Zitzler, E., Deb, K., Thiele, L.: Comparison of multiobjective evolutionary algorithms: empirical results. Evol. Comput. **8**(2), 173–195 (2000)
20. Czyzzak, P., Jaszkiewicz, A.: Pareto simulated annealing metaheuristic technique for multiple objective combinatorial optimization. J. Multi-Criteria Decis. Anal. **7**(1), 34–47 (1998)
21. Ulungu, E., Teghem, J., Ost, C.: Efficiency of interactive multi-objective simulated annealing through a case study. J. Oper. Res. Soc. **49**(10), 1044–1050 (1998)
22. Tan, K.C., Goh, C.K., Yang, Y., Lee, T.H.: Evolving better population distribution and exploration in evolutionary multi-objective optimization. Eur. J. Oper. Res. **171**(2), 463–495 (2006)

Index

© Springer Nature Switzerland AG 2021
M. Ahmadieh Khanesar et al., *Sliding-Mode Fuzzy Controllers*, Studies in Systems,
Decision and Control 357, https://doi.org/10.1007/978-3-030-69182-0

Printed in the United States
by Baker & Taylor Publisher Services